The Complete Guide to Successful Sprouting for Parrots

The Complete Guide to Successful Sprouting for Parrots

and Everyone Else in the Family

Leslie Morán

2007

Copyright © 2007 Leslie Morán
Critter Connection
PO Box 38
Silver Springs, NV 89429
www.moranscritterconnection.com
info @ moranscritterconnection.com

ISBN: 1-4196-8479-5
ISBN-13: 978-1419684791

The Complete Guide to Successful Sprouting for Parrots

Table of Contents

Foreword   ix

Introduction   xi

1. Why Feed Sprouts?   1

2. Organic or Commercial Foods, Is There a Difference?   13

3. Successful Sprouting   23

4. Caring for Your Sprouts   47

5. Now, Who Will Eat Them?   53

6. Nutritious Meals for the Whole Family   65

Appendix A — Resources and Additional Information   73

Appendix B — Trouble Shooting   83

Appendix C — Preventive Nutrition for Dogs and Cats   87

Appendix D — Homemade Egg Food   93

Appendix E — General Sprouting Guidelines   95

Bibliography   99

Periodicals   101

About the Author   103

Index   105

In Dedication: to Life, Health, and Happiness

Whether walking along a mountain trail, gazing out across the expansive Nevada desert, or running along the beach—I know I have been blessed. Experiencing and appreciating the beauty and complexity of this wondrous planet, nurtures me to the depth of my being.

I am grateful for everything that has illuminated my way, as I have grown towards following a natural and holistic path. For in having these experiences, I have been divinely guided to everyone who has contributed to making this book possible. From each person who read the manuscript, and helped with proof reading and editing, to everyone who helped me select the cover color and design. And even those who have been with me, solely, in spirit. I thank each one of you.

I also have immense gratitude for the synchronicity of life that drew towards me each situation that helped carry this book forward. I have felt myself being divinely supported, and I am thankful.

For my feathered family and my four legged critter kids, I have felt your unconditional love and encouragement daily. Thank you.

Special thanks to Michael. I have felt your nurturing support and I am grateful for your help—every step along the way.

Foreword

So often in life we discover that the simplest, most uncomplicated things are the most effective and useful. Those that cost little, and require little as far as time and resources are concerned, can often turn out to be a revelation in terms of their benefits—and so it is with sprouts.

In, *The Complete Guide to Successful Sprouting for Parrots*, Leslie Morán, with her dedication to happiness and well-being in both body and soul, brings to life the benefits and pleasures of including living sprouts in our diets—and those of our pets, as well.

Until fairly recently, these wonder vegetables haven't been given the recognition they deserve in modern, westernized diets, but the more one learns about them and how positively they can affect our health and general well-being, the more one wonders why we've neglected them for so long. A diet rich in fruit, nuts, seeds and raw vegetables is known to promote good health. Add to that sprouted, *living* vegetables, and we increase our chances of keeping tissues and cells clear of what could be called acidic debris, and of keeping our immune and digestive systems strong.

The most thought-provoking analogy I have come across is the comparison between a piece of cut fruit—which soon starts oxidizing and turning brown as it starts dying—and a jar of alfalfa sprouts which are still living and growing—full of life and goodness.

At a time when we're realizing how much damage is being done to our minds and bodies by careless eating and drinking, Leslie Morán's book comes along not a moment too soon. Leslie has researched her topic with the dedication and sincerity which she devotes to every subject on which she writes, and she presents her findings clearly and succinctly, in an easily acceptable style.

Gilly Lloyd
Editor, *Parrots* magazine

Introduction

Why I Sprout for My Birds

It was love at first sight when I decided to add my first birds, a pair of crimson-winged parrots, to my family. They had been fed a well known commercial pellet along with a dry seed parrot mix, had received regular additions of fresh fruit, vegetables, and a cooked legume and grain preparation. Even though many would consider this a balanced diet I wondered about some variations between the two birds.

Rainbow, the male, had radiant chartreuse green feathers. The plumage of his wings and tail were full and well shaped. His beak coloration was deep and rich. However, his mate Sunshine appeared different. Her feathers were drab and lacked a healthy sheen. The plumage of her wings and tail looked oily and thin, with the feather quills easily separating. By comparison the color of her beak was dull.

If had known then what I know now about the signs that indicate a healthy bird I might not have brought them home with me. But I was drawn to them and could not leave them where they were. As I considered the differences between the two birds I took a close look at what they were eating.

Since the mid 1980's, I have witnessed the results of the benefits of feeding fresh, homemade food to dogs and cats—and I've taught my clients and students about them too. When supplying our companion animals the highest quality of food possible the ingredients we use are equally as important as the foods we phase out and eliminate from their diet.

Everything an animal consumes must be processed by the body. The liver has a principle role in filtering out body wastes and any toxic chemicals that may be in our food, air, or water. These toxins, such as nicotine, environmental pollutants, and pesticides are excreted into the bile and are removed from the body.

However, with our environment becoming more polluted our food, air and water may contain more toxins than an individual's liver can properly handle. When this occurs the liver becomes over worked. This condition is described as having a sluggish or congested liver.

As I considered all of this I began searching for any information on holistic and natural care for parrots. Since a wholesome balanced diet is the cornerstone of good health I wanted to feed them a highly nutritious, high quality natural diet. I was looking for an avian diet that was comparable to the wholesome and health producing homemade recipes that I had fed my own dogs and cats for years. One that would help my first avian family members benefit from the highest quality nutrition available. I quickly discovered that there was little to no information available on holistic and natural care for parrots.

However, I did discover sprouting for birds. Then I remembered that in college I had learned about the nutritional benefits that people received from eating sprouts. Healing centers such as the Ann Wigmore Natural Health Institute and the Hippocrates Health Institute both teach and promote self-healing through using living foods. If eating sprouts and eliminating unhealthy foods helped restore health for people, could feeding sprouts help my parrots? Pieces of information began coming together. Gradually I began changing the diet Rainbow and Sunshine were eating.

I phased out the commercial pellets, that contained ethoxyquin a harmful chemical preservative that I knew stressed their liver, while I introduced an all organic pelleted food. I began growing and feeding them sprouted grains, seeds, and legumes. As they adapted I gradually removed the non-organic dry seed parrot mix and other components of their diet that did not meet the same quality of the fresh organic foods that I ate.

The seasons changed and both birds had been eating, and enjoying, their natural diet plan for about a year. I could see that my efforts had been richly rewarded. Both of them had developed plumage that glowed with vitality. All of their feathers were full and had a healthy sheen. The hen's beak had become a vibrant peach color more closely resembling the male's. These birds eagerly anticipated each sprout filled meal. Their eyes widening with recognition as they saw me bringing their fresh food to them each morning.

After placing their fresh food and water dishes in their enclosure, they quickly climbed down from their perches and began eating. They had developed wholesome appetites. Obviously they enjoyed their fresh food and found it much more delicious than their prior fare.

As my avian family grew my sprouting protocols also went through a metamorphosis. This flexibility ensures that all the birds under my care receive a nutritious and wholesome diet that is built on the foundation of feeding them fresh sprouts. This process has also allowed me to experience sprouting for a diversity of avian species—parrots, finches, quail, and other native wild birds—under a variety of conditions and circumstances.

A large part of the work I do with my clients is educating them in the methods and processes that I have found to effectively produce positive results with the animals. I have come to believe that growing sprouts for the birds under our care is the cornerstone to providing the highest quality of nutrition available. And in providing this high quality nutrition, good health, improved behavior, the healing of feather destruction, and radiant plumage all follow.

Although the primary focus of this book is to explain in detail how to grow sprouts for parrots and other birds, sprouts are also a delicious and nutritious addition to anyone's diet. This includes the family dog, cat, and even yourself.

As you read about the experiences others have had when converting their birds to a wholesome and natural diet that is centered around feeding freshly grown sprouts I hope you will be inspired to sprout for your birds. You will also discover how easy sprouts are to grow once you understand the proper process. Then you'll be able to begin reaping the benefits that others have received.

Good health to you and your birds, and have fun sprouting!

Leslie Morán

Chapter 1

Why Feed Sprouts?

"I want to provide the best nutrition possible for my birds", says Gail Worth, the owner of Aves International located in Rancho Palos Verdes, California. Three of her staff members provide 1,000 birds with sprouts daily.

In Waco, Texas, Jane Thomas at Hillside Aviary has been breeding birds for 25 years raising macaws, cockatoos, African greys, Amazons, cockatiels, budgerigars, ring-necked parakeets, and Senegals. In October of 2001 she converted her 65 birds to sprouts. Thomas says, "Feeding sprouts is more nutritious than cutting up fruits and vegetables, and it takes a lot less time." After only six months of having her birds on the sprout-based diet she described their feather quality as, "Shimmering, they look like they're ceramic." She's convinced it's the only way to go.

Even avian rescue organizations have recognized the benefits of feeding sprouts to the birds under their care. Jill Bell, at The National Parrot Preservation and Rescue Foundation in Humble, Texas, says that, "The biggest change in feather condition I have ever seen in my birds was the first complete molt after I started seriously feeding sprouts. And I still believe that today—many years later. I am a tremendous proponent of sprouting, as you can tell! "

What led Worth to convert her extensive aviary and why did Thomas change practices that she had employed for 25 years? And how was Bell convinced to feed sprouts to the birds under her care?

Their Superior Nutrient Content

When any nut, seed, legume, or grain is sprouted the chemical makeup changes. The sprout has two unique qualities. First, it is the only food that is fresh up until the moment it is eaten. And secondly, because it is a living food it contains life force energy.

Sproutable foods have between seven and 40 per cent protein. According to Brian Clement in *Living Foods for Optimum Health* during germination starches are converted into simple sugars, protein chains are broken down into their basic amino acids, fats are converted into soluble fatty acids, and vitamins are produced.

Sprouts are a rich source of vitamins A (beta-carotene), C, E, B, and antioxidants. In oats vitamin C increases 600 per cent after sprouting. Several sources including, *Handbook of the Nutritional Content of Foods*, prepared by the USDA state that while dry seeds, grains, and legumes are rich in protein and complex carbohydrates—they completely lack vitamin C. However after sprouting their vitamin C levels escalate to approximately 20 milligrams (mgs) per 3.5 ounces.

At Yale University Dr. Paul Burkholder studied the nutritional value of sprouted oats. His research determined that sprouted oats contained 10 percent more thiamine (vitamin B1), 1300 percent more riboflavin (vitamin B2), that pantothenic acid (vitamin B5) had increased 200 percent, pyridoxine (vitamin B6) had multiplied 500 percent, and biotin levels had raised by 50 percent.

Living foods that have been sprouted contain abundant amounts of natural minerals. During the sprouting process the minerals present become chelated, which means that they combine with other molecules in a way that makes them easier to assimilate.

William Peavy, MA, Ph.D., and author of *Super Nutrition Gardening* has pioneered the scientific study of the nutritional importance of germinated seeds for over 40 years. Peavy asserts that during the sprouting process enzyme levels rise. These food enzymes readily digest proteins and carbohydrates making the protein in sprouts more easily digested and assimilated by the body. This increased volume of enzymes also means that more are available to be consumed as biological catalysts that nurture and support every single physiological process in the body.

According to Phyllis Balch, CNC, in her book *Prescription for Nutritional Healing* sprouts are the richest source of enzymes available. Even though all raw foods contain enzymes the quantities that sprouts contain can be 10 to 100 times higher than in fruits and vegetables. Enzymes have a vital role in sustaining life. Balch affirms that even in the presence of sufficient amounts of vitamins, minerals, water and other nutrients life—as we know it—could not exist without the presence and action of enzymes.

Enzymes play an active role in virtually all of the body's biochemical processes. They are essential for digestion, proper functioning of the brain, provide the cells of the body energy, and are instrumental in repairing and regenerating every tissue, organ, and cell in the body.

The Value of Proteins

Like enzymes, proteins are also essential to life. Every living organism—massive or microscopic—is composed of protein. It forms the structural basis of chromosomes. The genetic code contained in each DNA strand is the information for making that cell's unique protein chain. Proteins are the primary building blocks for muscles, blood, skin, feathers, nails, and vital internal organs. They are also essential for proper growth and development. The body's ability to form and regulate hormones, enzymes, antibodies, and to properly process the elimination of bodily wastes all occur because proteins are present. Proteins are responsible for every life sustaining biochemical process in the body.

Amino acids are the building blocks of proteins. They are grouped into two classes, essential and nonessential amino acids. Essential amino acids must be consumed in the diet. The nonessential ones can be synthesized by the body if all the necessary nutrients are available in the proper quantities.

Protein Rich Foods

Foods that contain protein are divided into two categories, complete or incomplete proteins. Complete proteins contain abundant amounts of all the essential amino acids. They are found in meat, fish, poultry, eggs, cheese, and milk. However, these proteins are more difficult for the body to digest and assimilate when compared to the protein present in sprouts.

The other category of protein foods only contain some of the essential amino acids. Because of this we call them incomplete protein foods. These proteins are found in seeds, grains, beans, legumes, and an assortment of fresh vegetables.

No Dairy for Birds

Before continuing, it is important to note:

A PARROT'S DIGESTIVE SYSTEM DOES NOT HAVE THE DIGESTIVE ENZYME LACTASE. LACTASE IS REQUIRED TO DIGEST THE LACTOSE IN CHEESE OR DAIRY PRODUCTS.

Because of this we must cross both cheese and all dairy products off of our list of complete protein foods for birds. Feeding a food an individual cannot digest is one way of inviting digestive difficulties, which can lead to health problems, and can create unnecessary emotional stress due to the troubling physical discomfort.

Essential Amino Acids

For humans there are nine essential amino acids. These are lysine, methionine, tryptophan, threonine, leucine, isoleucine, histidine, valine, and phenylalanine. Birds have ten essential amino acids that must be included in the diet. The avian essential amino acids include the nine listed above, plus arginine.

Many of these essential amino acids are present in a wide variety of foods. However, four of these essential amino acids: isoleucine, lysine, tryptophan, and methionine are only present in sufficient quantities in certain foods.

To further clarify this we can divide the list of incomplete protein foods that contain these four essential amino acids into two groups. The first consists of legumes: beans, peas, and lentils. These foods contain high amounts of isoleucine and lysine, but are deficient in sufficient amounts of tryptophan and methionine. Soy, navy, pea, and white beans are also deficient in valine.

The second group of foods consists of nuts, seeds, cereals, and grains. These foods contain high amounts of the essential amino acids tryptophan and methionine, but are deficient in sufficient amounts of isoleucine and lysine. Peanuts and pistachios are also deficient in threonine.

Seeds—A Protein Deficient Diet

This straight forward yet often overlooked fact regarding the nutritional importance of having all the essential amino acids present illustrates the basic and indisputable fact that parrots and other birds being feed solely a dry seed diet are being fed a protein deficient diet. These birds will eventually succumb to the effects of severe malnutrition. Seeds are an incomplete protein source because they lack sufficient amounts of the two essential amino acids isoleucine and lysine. Due to this nutritional deficiency dry seeds cannot meet the protein requirements and nutritional demands of the avian body. Since amino acids are the building blocks of protein that are essential to life consuming a balance of the essential amino acids is vital for every life sustaining process in the body.

Even after decades of avian husbandry malnutrition is still the leading cause of illness, disease, and early death for companion birds. This statistic is heartbreaking primarily because malnutrition can so easily be prevented.

The information presented in this book empowers you to help make a difference. This book provides you the resources to ensure that all the birds under your care—from this day forward—can receive a well-balanced, nutritious, complete protein diet. You can now also share these nutritional facts with others you know who love and care for avian companions.

The Value of Proper Food Combining

In 1971 Frances Moore Lappé wrote her landmark book, *Diet for a Small Planet*. Still in print today, this book brought attention to the significance of proper food combining by assuring all the essential amino acids are present in a meal when eating a plant based diet.

In bringing the topic of proper food combining into avian nutrition being able to provide a wholesome and balanced complete protein diet for birds is really very easy.

As Lappé points out in her book human cultures all around the world spontaneously developed plant based ethnic foods that provide complementary proteins. For example Mexican food combines beans and rice, while Eastern Indian cuisine combine lentils and rice. Both beans and lentils contain high amounts of isoleucine and lysine, but are deficient in sufficient amounts of tryptophan and methionine. While rice contains high amounts of the essential amino acids tryptophan and methionine, but is deficient in sufficient amounts of isoleucine and lysine.

The beans and lentils are rich in the essential amino acids that the rice lacks, while the rice is rich in the essential amino acids that the beans and lentils lack. When these two foods groups are eaten together the strength one group has in essential amino acids compensates for the deficiencies in the other. The result is that the body receives a complete protein plant based meal because when the foods from these two groups are eaten together all the essential amino acids are present.

An Exception to the Rule

According to several sources including nutritional studies done at Purdue University, West Lafayette, Indiana and the Department of Applied Nutrition and Food Chemistry at the University of Lund, Sweden, quinoa (pronounced keen-wa) contains a balanced profile of amino acids. Because all the essential amino acids are present in sufficient amounts quinoa is a grain that is a complete protein food. Unlike wheat, rice, or corn, quinoa does not need to be eaten in conjunction with complementary protein foods such as beans or legumes for it to be a complete protein.

Quinoa also has a relatively high calcium content when compared to other grains such as rice, corn, wheat, or oats. More on incorporating quinoa into a natural avian diet plan in the following chapters.

Sprouts for Improving Health

At Aves International Worth has been feeding her birds a blend of sprouts combines with a large variety of freshly chopped raw and cooked vegetables since 1998. She currently has African greys, Amazons, caiques, cockatoos, conures, eclectus, lories, macaws, pionus, and poicephalus parrots. Since this change she has seen fewer illnesses, less problems associated with egg production, and a higher quality of viable eggs. Her birds are healthier, have a deeper saturation of feather color, and maintain healthier weight levels. She has no obesity problems with her birds.

Thomas began feeding sprouts because she recognized the value of feeding life-giving foods. One of her cockatoos had a fatty lipoma near his cloaca which has gotten smaller since she converted her aviary to sprouts.

Fred Bauer developed his avian nutritional expertise over a period of 18 years of diet evolution with more than 400 birds of over 50 different species. During this time he observed that birds who typically developed signs of poor health were those that were fed ungerminated seeds. But these birds, and those with diseases described as incurable, he reports responded favorably when fed a diet of sprouts.

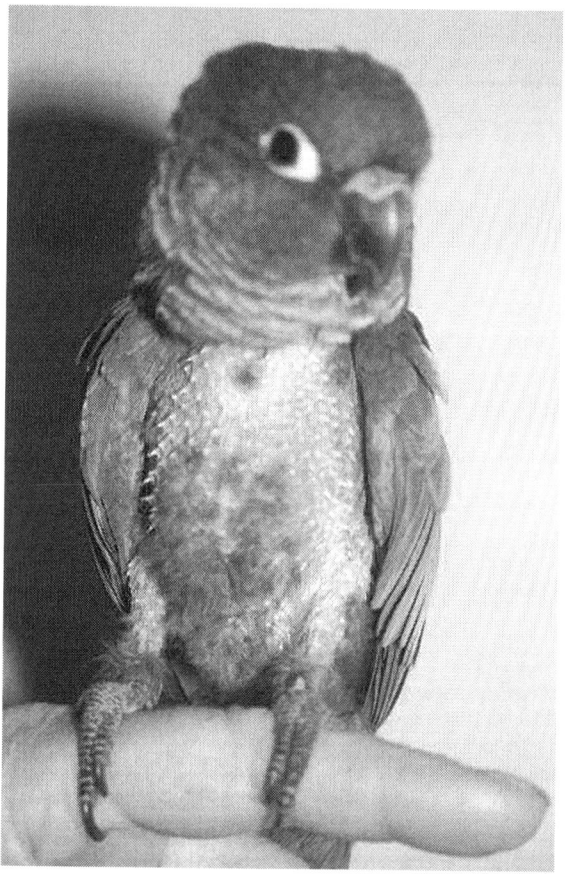

Elvis, green-cheeked conure, had plucked himself bald. Photo: Brice Clement.

Even birds who have developed feather destruction behaviors respond well to being fed a diet rich in a blend of sprouted legumes, seeds, and grains. Although healing feather destruction is a very involved situation a powerful foundation for reversing this condition can be laid by feeding a balanced blend of organic sprouted legumes, seeds, and grains.

An Individual's Response

Healing is a very personal journey. When holistic and natural methods are used to restore health and wellness the body's own healing and regenerative processes are influenced in a completely different manner than when a more traditional health care approach is used. Pharmaceutical drugs produce changes, but they can never heal. Only the body can do this. And the body can only heal and regenerate itself when it has everything it needs in sufficient amounts to perform this function properly.

When utilizing a natural approach to health and wellness the causes underlying the reason the health condition or imbalance appeared in the first place can often be addressed on a very basic level.

When providing the body the nutrition it needs whole, balanced foods, including complementary protein sprout mixtures fed with fresh raw fruit or raw or cooked vegetables, and species appropriate nuts provide an exceptional dietary foundation. Other nutrients could be in the form of nutritional or medicinal herbs, specific nutritional supplements: vitamins, minerals, antioxidants, or metabolic enhancers such as DMG (dimethylglycine). Individual nutrients differ in form and function and in the amounts needed by the body. And each individual person or animal has unique nutritional needs that are as distinctive as their personality.

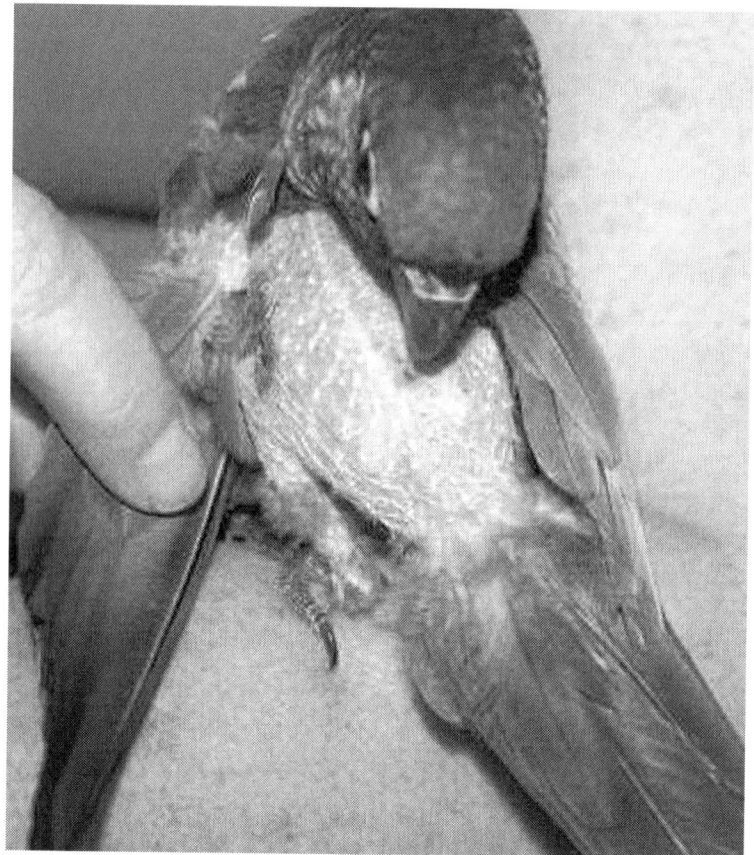

Elvis, green-cheeked conure, had plucked everywhere he could reach. Photo: Brice Clement.

As the body receives the type of nutritional support it needs regeneration begins at the cellular level. This process naturally takes time. As old cells die they are replaced by new healthier cells. This process is a basic principle of nutrition and is called physiological dynamics.

In people a blood cell lives between 60 to 120 days. In three to four months a person's entire blood supply has been completely replaced. In six months nearly all the proteins in the body have been replaced, including the DNA of the body's genes. In a year all of the bones in the human body have been replaced. Although these statistics have been documented in human healing and regeneration processes, be assured a similar regeneration pattern exists for parrots, our other companion animals, and wild animal species.

After starting any nutritionally based healing program expect a period of time to pass before you begin seeing improvements. Developing an attitude of patience with the natural restorative processes of the body is a way of respecting the innate healing wisdom within—the intelligence present at the cellular level. As you cultivate this perspective you also honor the sanctity of this wondrous healing and regenerative system that only exists in a living being.

Only the Best

With sprouts providing a wide array of nutrients that are easy to assimilate, rich in life enriching enzymes, and packed with increased vitamins, and antioxidants it's easy to under how an individual's health can improve. And for our avian companions with improved health comes beautiful feathers radiating a rich color that can only be found on a truly healthy bird.

By selecting ingredients that constitute a sprouting mixture representing proper food combining, highly digestible, and easily assimilated complete vegetable proteins can be readily grown at home right in your kitchen. Before covering the basics of growing sprouts the next chapter discusses the importance of using the highest quality ingredients available for your sprouting mixtures.

Elvis fully feathered. This green-cheeked conure responded very well to a holistic approach. Sprouts formed the basis of his personalized nutritional and healing plan.

Chapter 2

Organic or Commercial, Is There a Difference?

When grocery shopping for yourself and your parrots have you wondered about the nutritional value of the fresh fruits and vegetables you're purchasing? Why do organic grains, nuts, seeds, and legumes cost more? Are the pesticide residues in commercially grown foods really worth avoiding? Does your bird love to crack open raw peanuts in the shell? If so, did you know that if those peanuts were grown commercially they have a higher potential for absorbing the chemical fertilizers, soil fumigants, fungicides, and pesticides used during its growing season because peanuts grow underground?

Commercially grown fruits, vegetables, nuts, seeds, grains, and legumes routinely have an overwhelming volume of chemicals used on them on a regular and ongoing basis. These synthetic compounds inundate the plants and penetrate the soil they are grown in. Products used include herbicides, fungicides, and soil fumigants that sterilize the ground and pesticides that indiscriminately kills all insects—even the beneficial ones.

A report in the April 1988 issue of *Food Science and Technology Abstracts* stated that apples sprayed with pesticides retained 50 to 100 percent of chemical contaminant residues, even after several months in cold storage and a detergent wash.

What is Organic?

When distinguishing commercially grown foods from organic produce, strict growing requirements must be met and maintained. Currently, the standards for organic food vary from country to country and state to state. Read the ingredient labels on all food containers. Foods labeled 'no spray', 'pesticide free', or 'residue free' are not organically grown. Artificial colorings, flavor enhancers, or chemical preservatives, such as ethoxyquin, should not be used in any packaged food labeled 'organic'.

Because these discrepancies exist it would be wise to familiarize yourself with the specific standards for organic foods in your region. The International Federation of Organic Agriculture Movements (IFOAM) has established strict guidelines for its members to follow. You can view their information detailing the development of organic agriculture around the world on their website. The goal of this multinational, nonprofit organization is the worldwide adoption of ecologically, socially, and economically sound agriculture systems that are based on the principles of organic farming practices.

Organic agriculture involves more than growing food without synthetic fertilizers and chemical biocides. Growing practices include nourishing the soil by adding organic matter and other natural soil enhancing fertilizers that improve the fertility of the soil. This ensures that it is not depleted by repeated harvests and helps to provide for a sustainable future in agriculture. Global water sources become cleaner because less pesticides and other toxins find their way into surface water and aquifers. Organic farming encourages biodiversity. This benefits and helps improve the quality of natural ecosystems.

Biodiversity describes the variation, number, health, and diversity of different living organisms within a given ecosystem. Biodiversity is often used as a measure of the health of biological systems, ecosystems, and environments.

Global Statistics

The latest edition of *Organic Farming Worldwide 2007: Overview & Main Statistics*, edited by Minou Yussefi and Helga Willer, states that organic agriculture has been rapidly expanding and is now practiced in more than 120 countries around the world. According to their latest survey nearly 31 million hectares of agricultural and farm land are managed organically by at least 633,891 farms.

The figures for global organic agricultural, in descending order, are: Australia/Oceania: 11.8 million hectares, Europe: 6.9 million hectares, Latin America: 5.8 million hectares, Asia: 2.9 million hectares, North America: 2.2 million hectares, and Africa: .9 million hectares. These statistics do not include non-certified organic hectares that would raise these numbers even higher.

This study also documents the registered areas of organic wild collection lands. The largest collection areas exist in Europe and Africa and cover about 62 million hectares. The most important foods harvested in these areas are bamboo shoots, fruits, berries, and nuts.

When we add the registered organic wild collection areas to the global statistics the total figure rises to 93 million hectares of organic agricultural land around the world.

Organic Farming Worldwide 2007 indicates that the global interest in organically grown foods and products continues to rise. More and more consumers -primarily across North America and Europe—are realizing the personal and global significance of buying organic and they are voting with their money. In an effort to help meet the demands of this trend more farms around the world are converting to sustainable organic agriculture.

Are Organic Foods More Nutritious?

Let's consider how a plant grows. A seed is planted, watered, and as the ground warms from the sunshine, a sprout develops.

Roots tunnel downward and the cotyledons, the primary leaves of seed plants, begin reaching toward the sky. Aside from the water, this seedling absorbs all its nutrients from the soil it is growing in. Or does it?

Medical researchers Stephen Davies, MD, and Alan Stewart, MD, in their book, *Nutritional Medicine*, explain that the quality of conventionally grown food is so poor and is severely lacking in the nutrients it contains, that it can actually produce disease.

The nutritional value of a food is directly linked to the fertility of the soil. On an organic farm the soil has been carefully nurtured for years. Methods used that enhance natural fertility and biodiversity include crop rotation, planting nutrient rich cover crops that are turned back into the soil, and using organic animal and plant fertilizers and mulches.

If naturally occurring nutrients and micronutrients are not present in the soil, the plants cannot absorb them. Organic farmers and organic consumers acknowledge that the fertility of the soil directly affects the nutritional quality of the foods grown in it. After considering the care and feeding of the soil on organic farms, and how plants absorb these nutrients in the soil, it's easy to understand how the nutritional value of any organically grown crop far exceeds those produced by conventional farming methods.

Why They Cost More

Many aviculturists, bird lovers, and health conscious individuals choose to buy and feed organic because they want their birds and their family to be as healthy as possible. For those who have not yet made up their minds, understanding why organic foods are more expensive can help justify spending more money on them.

The production processes utilized in organic farming take more time, are more labor-intensive and generally have lower crop yields than conventional farming methods. However, government subsidies primarily support conventional farming methods by artificially lowering the prices of these foods. So are organic foods really more expensive? Or do conventionally grown foods cost less because they are available at prices far below their actual production costs?

If we take a look at the bigger picture, conventional farming does not include the price a person pays for veterinary or medical expenses when a family member succumbs to a disease or illness that could have been prevented by feeding a balanced organic diet. The price of foods grown with conventional farming also does not include expenses incurred from the need to clean up polluted water supplies. We often see these costs—that are directly linked to conventional farming—being subsidized through higher taxes or most costly water bills.

Organic farming practices directly address these major issues by providing safe, health enhancing, more nutritious food without putting more toxins into our already polluted environment. As more and more people buy organic the natural cycle of supply and demand will lower the cost of organic food.

Pesticides and Wild Birds

In the wild, birds have fallen victim to the pesticides in our environment. The decline of migratory songbirds was first documented in the 1940's. Then in the 1960's, predator birds—those living at the top of the food chain—such as the peregrine falcon, brown pelican, osprey, and the bald eagle— began experiencing reproductive failures. The chemical DDE, a compound produced when DDT degrades, caused the eggs of these birds to be laid with abnormally thin shells. These shells were so fragile that the eggs were crushed during incubation.

In 1995, over five per cent of the world's population of Swainson's hawks died in a heavily farmed region of Argentina. The pesticide monocrotophos, sprayed to kill grasshoppers, was found to be responsible for the deaths of these birds.

John Elliot, a researcher with the Canadian Wildlife Service, studies robins in the Okanagan fruit-growing region in British Columbia. These orchards have been sprayed with pesticides for over 40 years. Even though DDT was banned in 1972 the effects of this chemical continue to be documented. Elliot has found that levels of DDT in the birds' eggs are actually higher than the levels reported 20-25 years ago, suggesting that even though DDT is not currently in use, its chemical residue continues to affect these birds.

It is estimated that of the roughly 672 million wild birds exposed annually to pesticides on US agriculture lands, ten per cent are killed. A friend of mine lives near a prime California agriculture area. He told me, "After they spray the grapes near my community, there is a pile of dead birds at the end of the field. It's sickening."

Over the years, scientists from a variety of organizations have reported behavioral changes in wild birds contaminated with chemicals. In American gull and tern colonies in the Great Lakes, the Pacific Northwest, California and Massachusetts, field researchers have found nests with twice the normal number of eggs. This indicates that two hens are nesting instead of the expected male-female pair. In Lake Ontario, Canada, researchers noted that some gull and tern colonies displayed abnormal behaviors, such as less inclination to incubate their eggs or defend their nests. This leads to increased predation, and diminishes the hatching and survival of the chicks.

Even though research projects on wild parrot populations have not focused on the presence of pesticides, the observed effects on other species would allow speculation that it may only be a matter of time before abnormalities are observed. According to the Migratory Bird Center, at the Smithsonian National Zoological Park, "The full effect of pesticides on wild birds is extremely difficult to determine because most deaths go undetected".

Are People Impacted?

To date, the primary motivation for investigating the effects of pesticides on wild bird populations appears to be linked either to a massive die-off, a site in close proximity to human populations, or when the presence of pesticides threatens the economy. Research has shown that the chemical pesticides evaporate in warm conditions and settle in cool areas. They tend to accumulate in fatty tissues, have an accumulative effect on the body and can remain in there for years or decades before breaking down. Because of these qualities, it is estimated that every human carries several hundred synthetic chemicals in their body.

And Your Birds?

When reviewing the lengthy list of documented illness and diseases directly caused by pesticides, the primary information source comes from human based studies. According to the progressive report, *Pesticides and Human Health*, published in 2000, these include cancer, respiratory disease, neurological damage, reproductive and developmental problems, behavior abnormalities, and immune system damage.

This type of data does not exist specifically for parrots. Avian veterinarians and scientists specializing in this area of research refuse to subject parrots to a scientific study that they know would cause them to develop diseases already documented in humans and observed in wild bird populations. The current approach is to document the effects seen in parrots after feeding them a balanced organic diet, by taking note of the health improvements seen.

Greg Harrison, DVM, has been intimately involved in formulating and feeding parrots a balanced organic diet since 1985. In his practice, Harrison has seen the disappearance of cancer, fatty liver disease, and feather destruction in parrots converted to a balanced organic diet. According to Harrison,

"Pesticides act like the hormone estrogen and can damage reproductive organs and the liver."

Harrison reflects that many disease states have been prevented through feeding a balanced organic diet. Parrot breeders who follow his nutritional recommendations no longer see formerly common reproductive disorders such as overactive egg laying, egg binding, egg yolk sack peritonitis, or infertility. Other body systems strengthened by Harrison's preventive nutrition include the kidneys, lungs, and liver. He even suspects that hardening of the arteries (arteriosclerosis) in African Grey parrots is caused by the chemicals in their food.

At the 1995 American Holistic Veterinary Medical Association's annual conference, David McCluggage, DVM, offered nutritional recommendations for a variety of birds, including parrots. McCluggage emphasized the importance of feeding a variety of organically grown foods free of pesticides and artificial ingredients.

The Agency for Toxic Substances and Disease Registry (ATSDR) in the US issued a landmark report in 1997, which stated that a growing body of animal and human data suggests that many environmentally persistent compounds (pesticides and chemicals) have the potential to disrupt normal functions of the endocrine system. These chemicals may have a serious impact on reproductive and development parameters in wildlife and human populations.

The endocrine system of a parrot consists of the hypothalamus and pituitary gland, the gonads, pancreatic islet cells, adrenal glands, thyroid glands, parathyroid glands, ultimobranchial glands and the endocrine cells of the gut. This network of glands releases hormones into the blood stream,influencing countless bodily functions. These include growth and development, reproductive organ functioning, calcium metabolism, mineral regulation, feather formation, liver and kidney functioning and the secretions of other organs. A bird's metabolism is directly influenced by endocrine system hormones. Metabolism describes the sum of all chemical processes taking place in the body as they relate to the break down and digestion of food, resulting in growth, cell regeneration, use of energy, elimination of wastes and other bodily functions.

In essence, these environmentally persistent compounds present in our air, food, and water can adversely impact all of the body's biochemical processes essential to life.

Our Small Blue and Green Planet

For centuries, human civilizations have been impacted by epidemics. In the book, *Fowl! Bird Flu: It's Not What You Think*, the author, Sherri J. Tenpenny, DO (Doctor of Osteopathy), draws some fascinating conclusions about pandemics and pesticides. But what do these topics have to do with Avian Flu and the health of your parrots? At times, we live on a very small blue and green planet. The facts this book throws on the table, and weaves together with a compelling, fresh perspective, will astound you.

Fowl! cites chemical contamination of our environment, resulting in weakened immune systems, as being the primary cause of pandemics affecting people, poultry, and wild life. The information in this book has been extensively researched, and abundant footnotes are present throughout.

Fowl! also provides documented information that dispels the common misbelief that migratory birds transmit Avian Flu. Most importantly, outbreaks of this disease have not coincided with the arrival of migratory birds.

Any illness, whether in people, the wildlife, or our companion parrots, is tragic. No amount of justification can be made to try and compensate for the loss of life or the suffering endured. We can do nothing about what has already happened, however we can make an impact on the future. Knowledge is powerful, only if action is taken. This book substantiates the importance a single individual can make to help improve global conditions by making well-informed choices.

Why Buy Organic?

There are four primary reasons why people choose to buy organic foods, and these same four reasons apply when it comes to choosing organic foods for your parrots. Firstly, people who buy organically grown foods care about maintaining a high level of health and well-being. Secondly, avoiding produce and foods that contain pesticides and chemicals is an important part of preventive healthcare. What we choose to avoid is equally as important as the foods and practices we choose to use. Practicing this stage of preventive healthcare is significant in preventing the numerous diseases and illnesses directly linked to pesticide use. Thirdly, organic foods have a higher nutritional content. And, finally, pesticides are known to adversely impact our environment. This creates consequences that directly affect the parrots in our homes, in aviaries, and in the wild.

Our Global Community

Clearly, the decision to buy organic has personal and global repercussions. Awareness of the long-term deleterious effects that pesticides and other harmful chemicals have on living organisms, animals, people, and our environment is growing. Organizations such as Pesticide Action Network of North America (panna.org) are making powerful strides forward in helping to end the circle of poison. The 'circle of poison' describes the cycle of how pesticides that have been banned in the US, but are still being used in Third World countries, ultimately return to the US in foods and products imported.

As more and more people support a toxin free lifestyle by choosing to buy organic, we can and will see an end to the circle of poison. Will you ever look at a seed, nut, vegetable or piece of fruit the same again?

Chapter 3

Successful Sprouting

The practice of sprouting is more than twice as old as the Great Wall of China. In ancient Chinese manuscripts from 2939 BC, the country's emperor wrote about how good health came to those who ate sprouts. Today many people might enjoy bean sprouts in their Chinese food, or alfalfa sprouts in a sandwich or vegetable salad. But, this is just the beginning when it comes to sproutable foods.

When selecting foods for sprouting, seek out and use the highest quality ingredients possible. Avoid using chemically treated or preserved foods. Ensure they are being sold for human consumption, as many commercial seeds for gardening and farming have been coated with fungicides, herbicides, fertilizers, or growth hormones. Seeds treated with these chemicals can be poisonous if eaten.

Any chemical eaten by you or your birds will have to be processed by the body in some manner. The liver is the primary organ of detoxification in the body. It processes pollutants present in the food, water, or air. If an individual is exposed to more toxins than the liver can effectively filter, it becomes over worked. This condition is described as having a stressed or sluggish liver. When this occurs, you can provide some form of natural liver support. This will help it regain its ability to function efficiently. Birds are extremely delicate creatures and their sensitivity to chemicals in their food, water, or environment can vary greatly with each individual.

The highest quality ingredients you will find for sprouting will be organically grown, and can be found at natural food stores and from companies who sell foods especially for sprouting.

If you are ordering your seeds on-line or through the mail, be sure to specify you want untreated seeds suitable for growing sprouts for human consumption.

More About Quality

When shopping for sproutable foods, look at the individual pieces. There should be whole pieces of legumes, seeds, and grains. Broken fragments will not sprout, however they can cause spoiling to occur in the rest of your sprouting mixture. Sproutable foods should also be a healthy color. Mung beans are a deep green, brown rice is a light tan, and adzuki beans are a deep maroon.

Unhulled sunflower seeds are naturally a soft gray color. If you see any that are yellowish, they have become old, and many even be rancid. Foods become rancid because the oils they contain have become old and have gone bad. Rancid foods have a stale odor, like cooking oil that has spoiled.

Businesses who sell organic foods understand that freshness is an important quality. In addition to organic sprouting ingredients being free of pesticides and other harmful chemicals, they also possess more nutrients than their commercially grown counterparts. Typically, organic foods that can be used in sprouting will be fresh.

Dry Storage

When storing a sprouting blend, or ingredients you will use to make your own mixture, a cool dry location is preferred. Keep grain moths out by using an airtight zip lock bag, plastic jug, metal tin, or glass jar. Research indicates that sproutable foods can even withstand temperatures below freezing if these are above 10 degrees F, and they are double bagged.

Living Seeds

The primary reason for using high quality ingredients for your sprouting mixtures is freshness. In order for a seed, grain, nut, or legume to germinate, it must be viable. If any of the ingredients you purchase don't sprout, they may be old. And if they are old, they are no longer capable of providing the life enhancing qualities so vital to sprouted foods. In order to provide the life nurturing qualities that live foods are known, for they must still have that living germ—capable of growing—alive inside.

What Can Be Sprouted?

There is a long list of foods that can be sprouted for your parrots and yourself. Rather than covering all the exotic possibilities, this book will focus on a variety of foods that can be more easily obtained. These foods are also frequently used when sprouting for parrots.

You will want to refrain from sprouting any plants that have poisonous parts. Members of the nightshade family—potatoes and tomatoes—should be avoided. Potato sprouts, also called buds, are known to be poisonous and should not be eaten.

After you familiarize yourself with growing sprouts and have achieved a level of success, you may want to experiment with other less common sproutable foods. If I can't buy a sproutable food locally where I shop, I have omitted it from this book.

Foods that can be sprouted essentially fall into four categories: seeds, grains, legumes, and nuts.

Seeds you can sprout include fenugreek, flax, millet, pumpkin (also called pepitas), radish, sesame, safflower and sunflower (sprout alone).

Sproutable grains are amaranth, quinoa, unhulled barley (pearled barley will not sprout), buckwheat groats, brown rice, wheat berries, rye, and whole oats.

Legumes fall into three sizes tiny, medium, and large. The tiny ones are alfalfa and red clover. Although these are commonly referred to as seeds, their nutritional content clearly classifies them as members of the legume family. The medium sized ones are adzuki beans, mung beans, garbanzo (also called chick-peas), black turtle beans, and French lentils. The larger sized ones are soy beans and pintos beans. Raw unshelled almonds are a nut that also responds well to being sprouted.

Length of Germination Time

When considering which foods to blend together in a sprouting mixture, the length of time that it takes a food to germinate is important. If you combine quick sprouting quinoa with a slower growing mung bean, the quinoa can become extremely long before the mung will even begin to show a sprout bud.

The following sprouting time guidelines will vary from location to location, due to temperature and humidity.

In general, foods listed in the seed category will sprout and grow at a similar rate. However, it is suggested to sprout sunflower seeds alone. If unhulled sunflower seeds are soaked in the morning they can be ready to eat on your dinner salad that same day.

The sproutable grains listed previously are very similar in size and will also sprout and grow at a comparable pace. However, quinoa is another quick growing food. Because of this, it is suggested to grow quinoa sprouts alone, or with other tiny foods from the seed list.

Medium sized legumes can be sprouted along with any of the grains — except quinoa — or any of the seeds listed above. The large sized legumes do take longer to sprout, so it is suggested that a blend of these larger ingredients be soaked and sprouted as a mixture of their own.

Remember, these germination times are only intended as a guideline to help you determine which of these foods will

Foods that can be sprouted fall into four categories: seeds, grains, legumes, and nuts

- Seeds: fenugreek, flax, millet, pumpkin (also called pepitas), radish, sesame, safflower and sunflower (sprout alone).

- Grains: amaranth, quinoa, unhulled barley (pearled barley will not sprout), buckwheat groats, brown rice, wheat berries, rye, and whole oats.

- Legumes fall into three sizes: tiny, medium, and large.

  Tiny: alfalfa and red clover.

  Medium: adzuki beans, mung beans, garbanzo beans (also called chick-peas), black turtle beans, and French lentils.

  Large: soy and pinto beans.

- Nuts: raw unshelled almonds.

be a good sprout mixture for the growing conditions in your kitchen.

To help simplify things, you may want to purchase a prepackaged organic sprouting blend that can easily be shipped to you. Information in Appendix A at the rear of this book provides you this option.

Begin Where You Are

If you are presently feeding your parrots a dry seed mixture, this can possibly be sprouted. However, if it contains dry biscuit pieces, food pellets, or artificially colored vitamin pellets, the biscuits and pellets will turn to mush and the colorings will leach synthetic chemicals into the sprouts, contaminating them. These food items are not intended to be kept wet, and will grow mold and bacteria.

So, if you still want to sprout the dry seed mixture you already have, you'll need to pick out all the items that are not whole dry seeds. Depending upon your level of patience and your current seed blend, this may or may not be an easy task.

However, be aware that even after sprouting, seeds are still an incomplete source of protein. You will need to add some type of sprouted legumes to your bird's diet. Mung beans are often easy to locate, and all my parrots, from budgerigar to macaw, readily enjoy them.

Original Sprouting Blend

Our Original Sprouting Blend contains a balance of foods that provide a complementary protein mixture. The ingredients are French lentils, wheat berries, buckwheat groats, short grain brown rice, mung and adzuki beans. These ingredients will be appropriate for a variety of birds, ranging from parrotlets or small budgerigars to large macaws or cockatoos, and everyone else in between. For more information see Appendix A.

The Original Sprouting Blend contains a fairly equal

ratio of grain to legumes. These proportions help ensure that a healthy combination of all the essential amino acids are provided.

Mixtures for Small or Large Birds

Sprouts are a soft food that any curious bird, regardless of their beak size, will be able to eat. The Original Sprouting Blend provides a complementary protein balanced food for parrots of all species. The ones presently under my care range in size from a small budgerigar to blue and gold macaws. Although they all enjoy this sprouted blend, I also sprout a small seed blend for my budgerigar and finch flights. The ingredients of this mixture changes as needed; however millet is nearly always included.

For small parrots, budgerigars, cockatiels, finches, pheasants, doves, quail, and other birds more comfortable with small pieces, you can combine any of the listed seeds or grains with any of the medium sized legumes: adzuki beans, mung beans, black turtle beans, and French lentils. When creating your own sprouting blends, be sure to provide a fairly equal ratio of the seed or grain to the legumes. This helps provide complementary proteins in balanced proportions.

The Importance of Pure Water

Household water supplies generally come from surface water or ground water. Regardless of the source, our water is subject to contamination from a variety of impurities. It may contain substances and inorganic materials that the body cannot assimilate. And contaminated water can be one of the primary reasons that a person's early attempts at sprouting can fail.

Typically, the water that comes out of our faucets has been chemically treated in some manner. A range of chemicals, including chlorine, carbon, lime, phosphates, soda ash, and aluminum sulfate may have been added to kill bacteria, adjust pH, or eliminate cloudiness. Some localities may utilize more progressive water treatment methods, using ultraviolet treatment systems for disinfection or reverse osmosis filtration. However, each area is generally responsible for deciding how to process their water. Do you know how your drinking water is processed?

Whether or not you or your feathered companions are experiencing health challenges, I urge you to begin learning about your drinking water. After you understand where it comes from and how it is treated, consider improving the quality of it with your own water treatment unit. If your water comes from your local water company, they should be able to provide you with a copy of their water test report. This typically provides information on its source and how it is treated. If you have well water, Appendix A offers resources for having your water tested, and can help identify the best filtration system for your individual needs.

If after researching these options you determine that purchasing your own water filtration unit is impractical, using bottled spring water may be a viable alternative. However, you must also be diligent about ensuring that you are buying what is being advertised. In the US it is not uncommon for bottled water companies to treat water from the public supplies, bottle it, and declare it as being pure. A few years ago the media reported that a popular bottled drinking water was actually Detroit city water in an attractive container. Detroit, Michigan, has a population of over five million people and is a major industrial and automotive center. You can contact the bottled water company's manufacturer and request a copy of their water test report.

When considering the use of distilled water for drinking there is a fair amount of controversy surrounding this topic. Distillation is a water purification process where it is boiled, evaporated, and then this vapor is condensed. This process removes most of the bacteria, viruses, chemicals, minerals, and pollutants from the water.

Distilled water has been used to leach inorganic minerals from the body. Many question if distilled water also leaches important minerals, such as calcium, from the body as well. Some sources insist that this is not a concern, while others caution that this does occur.

Distilled water has also been called 'dead' because it lacks the life force nature puts into it. Natural fresh water sources contain minerals and electrolytes—both of these are necessary for good health. Since we want to eliminate all the toxins and contaminants in our drinking water, some suggest using distilled water and then adding organic minerals and electrolytes back in.

Although purchasing a home water distillation unit may be unreasonable for most people, you may consider having your water tested for certain pollutants and then use a filtration process that fits your budget and addresses our individual needs.

For sprouting, use the purest water possible. Many have been successful by filtering the chlorine out of their city's water supply and following the proper sprouting guidelines. However, if your sprouting attempts begin growing mold or have a bad odor, you may need to use purer water for your sprouting needs.

In Waco, Texas—which is hot and humid—Thomas says that water purity tops the list for successful sprouting. She uses distilled water.

Grapefruit Seed Extract—A Powerful Allay

One product that can help stop any bacteria or mold from growing on your sprouting mixtures is Grapefruit seed extract. Grapefruit seed extract (GSE) is a completely natural antibacterial, antiviral, antifungal, and antiparasitic product.

Grape seed extract is completely different from Grapefruit seed extract. Grape seed extract is an antioxidant that is found in the seeds of the white grape (*Vitis vinifera*), while grapefruit seed extract is a natural, non-toxic compound made from the seed and pulp of grapefruit (*Citrus paradisi*). This process converts the grapefruit bioflavonoids into an extremely potent substance.

According to research described in *The Healing Power of Grapefruit Seed* by Shalila Sharamon and Bodo J. Baginski, studies undertaken by a number of renowned institutes—including the Pasteur Institute in France and the USDA—have shown that that GSE is effective against approximately 800 bacteria and virus strains, 100 strains of fungus, and a large number of single cell parasites.

Additionally, this book cites that from 1991 to 1993, four US laboratories and one Canadian site conducted analyses on the efficacy of GSE. These studies concluded that GSE is effective in killing numerous gram negative and gram positive bacteria, fungus, and yeast including—but not limited to—staphylococcus, streptococcus, candida, salmonella, e-coli, and giardia.

Due to its natural antibacterial and antifungal properties, using GSE during the soaking process when sprouting for birds as a preventive measure, will stop bacteria and mold before it can even get started.

According to Thomas, if you live in a hot and humid area it is essential to use GSE during the soaking process. A proper amount of GSE in the soaking water helps eliminate the growth of any bacteria or fungus that may be present.

The antimicrobial properties of GSE occur at very low concentrations. For germinating seeds during the soaking process, we suggest using approximately 9—10 drops of GSE to 32 oz (1 liter) of water.

GSE has been effectively used as a nontoxic cleaning solution around the home, and can also be used for cleaning animal enclosures and cages. It is very effective in treating mildew. For information on obtaining GSE, see Appendix A.

Sprouting Supplies

Although there are companies that will happily sell you

plastic sprouting towers, containers, and buckets, many times the simplest supplies will work equally as well. But then the number of birds you will be growing sprouts for will ultimately be the deciding factor in determining the items you decide to use.

For sprouting in your kitchen, the easiest sprouting supplies to obtain are clean, wide mouthed glass jars, some nylon net, and rubber bands. For draining and growing sprouts in jars I find that a typical dish rack, set inside a square or oblong plastic tub, works very efficiently.

For draining and growing sprouts in jars a typical dish rack, set inside a square or oblong plastic tub, works very efficiently.

Mason canning jars with the two-piece lids are ideal for sprouting. Remove the center segment of the lid and use the screw-on ring to hold the nylon net in place. Other options are peanut butter or mayonnaise jars. If you're looking for a larger—half or one gallon size—jar, perhaps a delicatessen or restaurant in your area would be willing to save a few for you.

Nylon net can be purchased at fabric stores and is available in two mesh sizes. Hand wash and rinse it well before use. The regular nylon net has about an eighth of an inch opening and is fine to use for most sprouting foods. However, for a mixture that contains smaller food ingredients, such as millet, alfalfa seeds, amaranth, or quinoa, you will want to use bridal veil netting. It is a much finer mesh.

The regular size of nylon net can be folded double over the top of the jar and is easily held in place with a rubber band. When using the bridal netting, typically one layer works well. Because of the finer mesh, using two layers can make it difficult for the water to flow out during the rinsing process.

Air circulation is also an important component to successful sprouting. A double layer thickness of the regular sized nylon net allows for good air flow in and out of the sprouting jar. And a single thickness of the bridal veil netting will also allow for an ample air flow. However, two layers of the bridal veil netting can restrict the air flow enough to adversely affect the sprouts. But your best guide will be to try what is suggested and to learn what works the best for you.

The number of birds you are sprouting for will help you determine the size of jars you needs. I sprout for nine parrots under my care, and use one 48 oz and two 64 oz glass jars.

Supplies for Large Scale Sprouting

If you will be sprouting on a large scale, use five-gallon buckets with a vast number of tiny drain holes drilled in the bottom. Or you can soak your sprouting mixture in a pail or bucket and rinse them in a large metal colander.

Thomas in Waco, Texas, recycles two and a half gallon buckets that powdered milk comes in. After numerous tiny drain holes have been drilled in these buckets, they can be set inside a five gallon bucket for soaking. She then uses hospital grade cheesecloth (known as butter muslin to the British) and a very large rubber band to cover them.

If you are recycling any plastic or glass containers for sprouting, be sure they originally were used for the storage of human grade food products. Any plastic storage containers can absorb molecules of the items that have been stored in them. For this reason it is especially important to recycle and reuse only containers that have been used for storing human quality foods or beverages.

How Much to Sprout

When first beginning to sprout for your birds, it may take a few trials to determine how much of a sprouting mixture to sprout, how long it needs to soak, the length of time to germinate, and the number of days the sprouts must grow before they are at their optimal length.

Much of this data will be determined by the temperatures and humidity present where you live. In warmer climates foods will sprout more quickly, whereas in cooler areas your sprouting mixtures may need some additional warmth in order to properly germinate.

After the soaking process, your sprouting mixture can double in size. To prevent over flow, only fill your sprouting jars half full. This important fact can help you determine the size of jar to use, the amount of a mixture to sprout, and will help ensure that this mixture has been covered with a sufficient amount of water during the soaking process.

When determining how much of a sprouting mixture to germinate, consider the length of time they will need to grow. After soaking, it will take approximately three to four days for most of the sprouts to reach their optimal length.

Once you have grown your first batch of sprouts, you will

quickly learn how long this amount of sprouting mix will last. Then you can determine when you will need to begin soaking the next jar of sprouts to keep yourself and your birds well supplied.

While writing this chapter, I took a good look at how much I sprout, the number of parrots I feed sprouts, and the number of days the sprouts last. After doing the calculations, I arrived at one heaping tablespoon of the Original Sprouting Blend per bird per day.

For example: if you have four parrots and the sprouts will last three days, try soaking 12 tablespoons of the Original Sprouting Blend, and see if this amount is sufficient to feed your four birds for three days. And if it takes three days of growing time before the sprouts are ready, start soaking a new batch the day you begin using them. These figures are offered with the intention of giving you a starting point when deciding the volume of a sprouting mixture to sprout.

If you are primarily germinating tiny seeds like alfalfa, red clover, and radish, be aware that a fairly small amount can produce a large volume of sprouts. Experience has shown that approximately three tablespoons of these seeds will yield about four cups of sprouts.

For my finch flights, I soak between one quarter and half a cup of their seed mixture. After three days of growing, this amount will last between four to six days. Halfway through feeding these sprouts, I begin soaking another batch of seeds.

Soaking Water Temperature

When soaking your sprouting blend, try using warm water. This has a way of jump-starting the germination process. The moisture and the warmth are two key factors that encourage and promote a good germination response.

Put some of your sprouting mix into the jar and fill it about halfway with water. Then fill it the rest of the way with warm or hot water. The two temperatures will mix, resulting in a warm, soaking water that will accelerate germination. Do be aware that pouring hot water into a glass jar filled with only a dry sprouting mixture will shatter the jar. But blending warm and room temperature water can be done safely.

Length of Soaking Times

The length of time to soak your sprouting mixture will vary depending upon the temperatures and level of humidity where you are. In drier climates that are less humid, such as here in Nevada, sprouting mixtures can be soaked overnight for approximately 10 – 12 hours.

However, in Waco, Texas, Thomas reports that during times of the highest temperatures and body drenching humidity in her area, she can soak her sprouting mixtures for as little as six to eight hours.

Deciding upon the correct length of time to soak your sprouting mixture in your area will take a little computing. Compare the two climate extremes described above to where you live and make an estimate. Based on the results you see, decide to either lengthen or shorten the soaking time. This approach can be as simple and easy as you allow it to be.

Using GSE and 'The Sniff Test'

During the germination and sprouting process, natural changes will take place in the color and aroma of the foods being sprouted. Wheat can smell green, a little like cut grass. Grains and seeds can have a sweet fragrance, while some legumes can develop a starchy aroma, like fresh homemade mashed potatoes.

After rinsing the sprouts, I sometimes sniff them. If I detect a questionable odor, I will add a few drops of GSE and water to the jar, let them soak for a few minutes, and then rinse. Then I sniff them again. In many situations, soaking the sprouts in the GSE and water solution will neutralize whatever was causing the sprouts to begin to smell. However, if after using this soaking method the sprouts still have a doubtful odor, I throw them out and use another batch. Many times this occurs with the larger sized ingredients. These seem to need more frequent rinsing.

When working with your sprouting mixtures, if you notice they have developed a slimy appearance or have a rancid, fermenting odor—they have spoiled. Toss them out and begin again.

You can use 'The Sniff Test' at anytime during your sprouting process.

What Works For Me

I sprout for nine parrots using the Original Sprouting Blend. I use one 48 oz and two 64 oz (half gallon) glass jars. I put three cups of the Original Sprouting Blend in one of the 64 oz jars and cover it, using two layers of the regular sized nylon net, with thick rubber bands holding it in place. Next I rinse the sprouting mixture, to clean off any dirt or dust that may be on the ingredients.

I then put about ten drops of GSE on the nylon net and fill the jar about half full of filtered water. I place this jar inside a square plastic container, or bowl, to catch any accidental water overflow from the soaking process.

Heat encourages growth. To facilitate the germination process, I heat some water on the stove and pour it into the sprout soaking jar. This is when the square plastic tub or bowl, into which I placed the sprouting jar, catches any overflow.

As the seeds soak, they absorb water. When I notice the water level has decreased in the jar, I add more warm water.

I then allow the sprouting blend in the jar to soak overnight, approximately 10 hours.

Early the next morning, I begin rinsing the sprouting mixture. At this point I divide the now swollen sprouting mixture between the 48 oz and one of the 64 oz jars. I pour the liquid off, refill the jars and empty them two or three times.

I repeat this rinsing process as many times as it takes until the water being poured off looks clean and clear. If it appears cloudy—keep rinsing. Inadequate rinsing is another reason sprouts can become moldy or spoil.

In general, for the seeds to grow properly, they must be kept warm, rinsed daily, have good air circulation, and have light. Please don't try to sprout inside a cupboard or closet.

Next, the sprouting jars are placed—with the mouth facing downward—in the dish rack (inside the square plastic tub) to drain. The dish drying rack holds the jars at a 30 to 40 degree angle which facilitates air movement in and out if the jar. It also allows any excess water to drip out.

This sprout drain rack is placed in a warm location that receives natural light. Since there is a north facing window above my kitchen sink, setting the draining rack next to the sink works well.

Do not put your sprout jars in direct sunlight. The intense heat quickly burns tender shoots, reducing them to mush and making them unusable.

Because I live in Nevada, the humidity is fairly low, which means the air is dry. I rinse my sprouts at least twice a day, morning and evening. I fill the jars with water, drain it out and repeat this process at least two or three times. Sometimes, while rinsing, I shake the jars, which helps aerate the sprouts. Occasionally, I will rinse at midday, if it has been warmer than usual.

Now the sprouts grow. The temperature will dictate how quickly or slowly this occurs. During the summer, the kitchen stays between 70 and 90 degrees F. Even with a swamp cooler, when it's over 100 degrees F outside, an indoor temperature of 90 degrees F feels refreshingly cool. During the winter, I keep the house around 68 degrees F. This cooler temperature works well for the Original Sprouting Blend, but unhulled millet seeds need additional warmth, both for the germinating and growing process.

According to Peavy in *Super Nutrition Gardening*, sprouts that have been grown over a three to four day period reach their peak enzyme activity levels at this time. Typically it takes around three days for the Original Sprouting Blend to reach their optimum nutritional length in my kitchen year round. I begin using the sprouts in the 48 oz jar first.

However, my sprouting continues in an ongoing cycle. Let me explain. As my newly grown sprouts reach their proper length, I begin using them. After I empty the first jar, I now have two empty jars and one still containing healthy living sprouts. Since I am now halfway through my current supply of fresh sprouts, I soak another batch. Three more cups of the Original Sprouting Blend is poured into a 64 oz glass jar, and the overnight soaking process begins again.

Following the steps described above, as I rinse the newly germinated sprouts in both growing jars each morning and evening, I am finishing up the healthy living sprouts from the previous batch of sprouts. This ongoing cycle ensures that I always have sprouts available.

This section details the growing process. Caring for your living sprouts will be covered in the next chapter.

Exceptions to the General Sprouting Guidelines for Hot and Humid Areas

Heat, moisture and humidity all accelerate germination and the growing process. This section will make you aware of how to alter the general sprouting guideline to suit your special needs.

Jane Thomas in Waco, Texas, uses distilled water and GSE for soaking all her sprouting mixtures. She was taught to use two teaspoons (8 cc) of GSE in one gallon of water. This is a higher amount of GSE than people in less humid areas use.

Because of the warm temperatures and the increased humidity, she sometimes also reduces her soaking time, to as short as six hours. In her area, the summer humidity can range between 65 to 70 percent. During the winter it averages about 35 percent.

She is also adamant about NOT rinsing the sprouts. Because the sprouting mixture was soaked in a water and GSE blend, she thinks it is very important to leave any remaining GSE on the seeds, grains, and legumes as they germinate and grow. Because GSE is an all natural antifungal and antibacterial product, this helps prevent any microbial growth from occurring.

"When you have high humidity, your sprouts need lots of good air circulation, " Thomas affirms. Because she does not rinse her sprouts, she provides them air circulation by fluffing them. She does this by turning and flipping her growing buckets every time she walks by them.

To review, the three key factors Thomas describes that can ensure successful sprouting in a hot and humid area are, first—water purity. Use distilled water. Second, shorten the soaking time. Six to eight hours can be sufficient. And thirdly, provide lots of good air circulation by turning and fluffing your sprouts often. Doing this every hour or two would be ideal.

Special Treatment of Millet Seeds

When temperatures are warm, millet seeds will respond to soaking, germinating, and growing on the kitchen counter, right along with the Original Sprouting Blend. However, when fall transitions into winter, my kitchen counter becomes much too cool for the millet seeds.

To compensate for this, I use warm water and GSE during the soaking process. After rinsing them, I then place them in their own small square plastic container that holds the jar at the proper angle. Next I place this on the top rack of the oven. The heat from the gas pilot light keeps the inside of the oven at a toasty 85 degrees F.

Yes, the general sprouting guidelines do explain that sprouts need light to grow. However, millet seeds apparently require warmth more than light.

Before I started placing the millet seeds in the oven, they

were not germinating and would spoil. Sometimes a healthy seed just needs a little heat to spring to life.

In comparison, the Original Sprouting Blend I sprout year round on my kitchen counter germinates and grows very well at the lower winter temperatures which run between 65 and 68 degrees F.

When germinating and growing millet seeds in the oven, be sure to rinse them at least twice a day, morning and evening. If you forget, they can quickly begin growing mold. If this happens, toss them out and begin again.

Sprouting Sunflower Seeds

This is another seed that really likes being warm before germinating. There are three different ways you can sprout them. Ideally, you can achieve the best results by growing them alone. If sunflower seeds in the shell are a part of the mixture you are trying to sprout, and they do not receive the warmth they need, they can spoil and would taint the rest of your sprouting mixture.

The first method is the most reliable. Buy the raw, unsalted, unshelled sunflower seeds. These will germinate very easily and quickly. If you look closely after soaking them for 10 to 12 hours, you will observe that the tip has grown ever so slightly. If you soak them in water at bedtime, they will be ready to feed for your parrot's morning meal. Conversely, if you soak some in the morning, they will be ready for you to toss on top of your green vegetable salad at dinner time. They have a mild nutty flavor and can be enjoyed by both you and your birds.

If these sprouts are allowed to grow any longer they can develop a bitter taste, so I prefer eating and feeding them when they are short. However, you could also try growing them a little longer, using the rinse and drain process described, to see which length your birds prefer.

The second method is less dependable, but it may work well for you. For a very long while I was able to buy the raw sunflower seeds in the shell, soak them, and they would germinate and grow on the kitchen counter. Then when the winter temperatures became prevalent, they stopped germinating and would spoil.

So, I moved them into the oven and used the process outlined above for millet. And this process worked for a while. But then all of a sudden, the organic sunflower seeds in the shell would not grow at all. When this happened, I changed to using the unhulled seeds, and the birds didn't seem to mind having them already shelled for them. When you can get sunflower seeds in the shell to germinate, it is best to feed them when they are short, otherwise they can become bitter. If a sunflower seed in the shell needs temperatures higher than 85 degrees F to germinate, then proceed to the final way of sprouting them.

The third method is to grow sunflower greens. Grab a flower pot made of clay, pottery, or plastic. Add potting soil, plant the sunflower seeds about a half-inch deep. Soak well with water and place in the direct sunshine. Heat and warmth are a vital part of growing sunflower greens. Within two to three days, the seeds will germinate, and delicate sunflower greens will push their way up through the soil. Ideally, you want to harvest them soon after they push through the soil, otherwise they will get a little tough. These are also delicious to add to your dinner salad, and make an interesting decoration on top of your bird's fresh food. Once your birds taste them, you may need to grow them on a regular basis.

You can harvest your sunflower greens and store them in a plastic bag in the refrigerator. They will keep for a few days, until you can ready your next batch for harvest.

Sprouting Quinoa

This is another food that is best grown alone. Like unhulled sunflower seeds, quinoa also germinates very quickly. After soaking them overnight on my kitchen counter, I find that they have already begun sprouting. Usually, by the second morning, they are ready to eat.

Gail Worth, at Aves International, told me that quinoa will

sprout for her in an hour and a half. Now considering she is in southern California, the warmer temperatures clearly accelerate this grain's ability to germinate and grow quickly.

Quinoa can also successfully be combined with any of the small seeds such as: alfalfa, red clover, or flax. Experiment to discover the combination that works the best for you.

Garbanzo Beans

Other legumes that are comparable in size to this bean will take a few days to germinate and begin growing. Garbanzo beans, also called chick-peas, are another food that can be ready to eat overnight. After soaking this bean for 10 hours, many times you will notice that its tip has become elongated. You can sample this day-old sprout to check for tenderness, or you can follow the rinsing and draining process for two to three days, allowing it to grow a little longer.

Soaking Almonds

According to Peavy in *Super Nutrition Gardening,* there are some benefits to soaking raw almonds overnight before eating them. After being soaked, they swell, which he describes as a presprouting process that activates the living germ inside. As with other sprouts, soaking almonds increases their enzyme levels as they become tender. Whenever I offer soaked almonds to my parrots, they savor them as if they are a special treat.

Amaranth

After germination, the amaranth sprout becomes a pinkish-red. Other seeds of a similar size, such as alfalfa or red clover, do not develop this characteristic. However, this blush of color is normal for this seed.

What Works for You

Once you determine which sprouting methods will work the best for your individual situation, you can quickly find yourself with a rich bounty of delicious sprouts. After growing them, the next stage is learning how to care for them. Since they are a living food, they will need a little bit of TLC each day. The next chapter will direct you in keeping them fresh and vital.

Chapter 4

Caring for Your Sprouts

Fresh living foods, vital and full of enhanced nutrition, need some simple and basic care. This ensures that they remain healthy and viable while you are feeding them to your birds, and enjoying them yourself.

The storage methods described in this chapter can be used for the Original Sprouting Blend, or any other mixture of seeds, grains, and legumes you combine and sprout. The foods with special sprouting needs covered in chapter 3, are discussed near the end of this chapter.

In learning to care for your sprouts, it is important for you to provide the three things that you supplied during their growing cycle. They will still need appropriate temperatures, rinsing, and adequate air circulation.

However, because the focus is now on proper storage instead of germination and growth, there will be a few changes to the routine.

Up until now, the guidelines have instructed you to keep them warm so they would transform from dormant seeds to living sprouts. Now cooler temperatures will help keep them fresh in storage while they are being eaten over the next few days.

The best way to provide lower temperatures is to store them in the refrigerator. The cooler temperatures do not stop the sprouts from growing, but it does slow their growth significantly.

Although many might want to shovel their sprouts into plastic bags, this is not recommended—for two reasons. First, sprouts are alive, and living plants breathe. Even when being kept in the refrigerator, they will need to have adequate air circulation. Being stored in a plastic bag does not provide this and would cause them to become slimy and spoil before they could be used.

Secondly, in order for your sprouts to remain as fresh as possible, they will still need to be rinsed. And an important part of the rinsing process is providing adequate drainage. If they become water logged, or are stored with an excess of moisture, this can also contribute to them becoming slimy or mushy. If this occurs, they would need to be thrown out. Plastic bags are not designed for rinsing and draining sprouts. But the jars you already have them growing in are ideally suited to this purpose.

You have already developed a rhythm of rinsing, draining, and providing your sprouts proper air circulation during their growing cycle. Continuing these practices while they are being stored will keep them fresh and vital. Ensuring that your sprouts still receive the proper daily care they need, will help them maintain the highest nutritional levels possible. This will also prevent them from spoiling before they are used.

Proper Storage

The method described here will enable you to provide your sprouts with everything they need to remain fresh and delicious for as long as possible, while they are being stored and eaten.

When your sprouts have grown to their optimal nutritional length, which is generally three days, you will want to rinse them well before using. After spooning out the amount you need, store the remaining sprouts in the refrigerator.

Since you are in the habit of rinsing your growing sprouts and setting them to drain at a 30 to 40 degree angle, you will want to move this idea into the refrigerator. This is the best way to ensure that they will have adequate air circulation while being stored. It is also a fairly easy task.

Recover the jars with the nylon netting held in place with a rubber band. Locate some small to medium sized square, round, or oblong plastic containers that have sides about two inches tall. Prop each sprouting jar inside a plastic container at an angle that allows any remaining rinse water inside to drain out. Then place your sprouts jars to drain in the refrigerator. If water has collected in these plastic containers, be sure to pour it out.

Locate a small to medium sized plastic container that has sides about two inches tall. Prop a sprout jar inside at an angle.

The next time you need to use some sprouts, bring them out, rinse and drain each jar. Spoon out the amount you need and replace the remaining sprouts back in the refrigerator to drain.

Even if you will not be using a particular jar of sprouts, they still need to be rinsed and drained at least once a day while being stored in the refrigerator.

GSE and 'The Sniff Test'

Healthy growing sprouts can develop an array of scents. Wheat can smell green, a little like cut grass. Grains and seeds can have a sweet fragrance, while some legumes can develop a starchy aroma like fresh homemade mashed potatoes.

Any time while storing your sprouts, you can sniff them to ensure they are still in good condition. If they smell off, add some GSE and let them soak for a few minutes. Then rinse and drain repeatedly, and sniff them again. If they still don't smell right, don't take any chances, toss them.

During storage, the most common time you may need to do this is when you are nearing the end of a batch of sprouts. Occasionally they just need a brief soaking with GSE to help them supply the last serving of sprouts for either you or your birds.

Then place your sprouts to drain in the refrigerator.

Storage That Works For Me

After following the growing guidelines detailed in the last chapter, the two jars filled with sprouts made from the Original Sprouting Blend will last me and my birds for about six days.

While using the storage methods described above, rinsing and draining both of the jars everyday is one of the key reasons that they will stay fresh and vital for this length of time.

Storage in Hot and Humid Areas

Returning to the high humidity in Waco, Texas, Thomas explains that she begins soaking a new batch of sprouts every other day. She has arranged her sprouting schedule so that her supply of sprouts lasts for two days.

After feeding her birds, she will store the remaining sprouts by pouring them into a very large bowl. Recognizing that they will still need to breathe, she covers the bowl with a single layer of dry paper towels. These are held in place by a fine screen that is placed on top of the bowl.

Remember that due to the high humidity in her area, Thomas does not rinse her sprouts. Because of this, they still have the GSE on them. This means that the following morning, she simply scoops the sprouts out of the bowl and feeds her birds.

Millet, Sunflower, Quinoa, and Garbanzo Sprouts

Although these foods have special growing needs once they reach the desired sprout length, they can easily be rinsed, drained, and stored in the refrigerator, using the same guidelines described above for the Original Sprouting Blend and other sprout mixtures.

Sunflower Greens

Sprouted seeds, grains, and legumes thrive from being rinsed, drained, and stored in the refrigerator, using the jars they were sprouted in. However, sunflower greens do not require the same level of care.

When sunflower greens are harvested and cut from the flower pots they grew in, they can be placed in a plastic bag for storage in the refrigerator. You can experiment with the best time to rinse them. Depending on the level of humidity, or lack of it, in your area, there are two options. First, you can rinse them before bagging. Include a dry paper towel to absorb any excess water. Watch them to ensure they don't become slimy. Or secondly, you can harvest them and bag them dry. Then you can rinse them right before you feed them to your parrots or toss them on top of your salad.

Almonds

After your almonds have been soaked, and they are nice and plump, it's often a good idea to rinse, drain, and refrigerate them for storage in their soaking jar. This allows for good air circulation. Since they are ready to eat in such a short time, make an effort to use them up within a couple of days. Or you can soak your raw almonds overnight for use the following day.

Ready to Eat

Beautiful, healthy living sprouts are such a healthful addition to any meal. Now that you have them, the next chapter describes how to ensure they are eaten and enjoyed!

Chapter 5

Now, Who Will Eat Them?

Experience has shown that between 70 and 90 percent of the birds introduced to sprouts enthusiastically eat them the very first time. When I introduced sprouts to my flock, nine out of 13 birds were in this category. These included two out of three budgerigars, a pair of scarlet-chested parakeets, a pair of Amboina king parrots, one of two blue and gold macaws, and a green-cheek and white-eyed conure.

Worth at Aves International says, "You have to treat birds like children—make sure they eat their broccoli. It's easier to convert a bird to eating sprouts than people think it is. Often times it's the people who are concerned about trying something new. Whenever I acquire a new bird, he is fed what all my other birds eat—a diet that is as close as we can get to what they would eat in nature—and I've never had a bird refuse to eat."

Expectations

Worth is right. Often times it is a person's opinion or attitude that influences the behavior of a parrot or other companion animal under their care. Why do some birds readily accept sprouts being introduced into their diet while others do not?

Several factors influence this condition. In my experience, animals—and especially parrots—are quite capable of making decisions based on prior experiences they have had. They do this in a manner much like people do. And, like us, the animals then make conscious choices regarding a variety of situations.

So, based on a parrot's expectations, and previous experiences, he will have beliefs and behavior patterns that have developed from these encounters. Because of this, each individual will have their own range of possible behaviors when responding to a given situation.

What are your expectations about how your bird, dog, or cat will respond to something new?

For a very long time, information has been available regarding how strongly our thoughts, expectations, and beliefs actually affect and create the experiences we have. In understanding this phenomenon, we have come a long way from Norman Vincent Peale and the power of positive thinking. So how does this apply to helping your birds readily accept the addition of sprouts to their diet?

When we think, we automatically form pictures in our mind's eye. If you're reviewing how your greenwing macaw, Sam, always tosses any new food item across the room, as you think these thoughts, you have a very detailed action film running through your mind. Animals see these mental pictures we inherently create and act upon them. Simply because we think about a behavior our bird, dog, or cat has done, he automatically thinks we want him to do more of that particular behavior. Why? Because we're concentrating on it, and what we focus on expands.

When working to improve the quality of a behavior, especially when introducing sprouts into the diet, always focus on and think about the result you actually want to see occur. If you catch your thoughts wandering, take a breath and refocus.

Review the image in your mind of Sam curiously investigating the sprouts in his food dish. Envision his response as he tastes one. See his eyes pin, meaning, "WOW this is good". Hear him say, "Mmmm!" Expect to walk by the food dish later that day and see it empty—every sprout having been eaten.

If you find your thoughts returning to old, unwanted,

behaviors, take a breath and regroup. When learning to develop any new habit, it takes patience, practice, and persistence.

Even complaining, or worrying about, unwanted behaviors also funnels energy towards having them continue. Put a smile on your lips, feel it on your face, and think about the behavior you want to create. Thoughts have energy, and this energy creates our reality.

An Animal's Learning Process

Previous experiences our animals have had is one way that they learn. The familiarity of being encouraged to investigate new things creates an attitude of feeling safe while exploring something never seen before.

But what if your bird has developed a different belief system surrounding feeling safe while inspecting something unfamiliar? Then you want to help encourage him to adopt some new habits.

Before Changing a Thing

Before making any changes to your bird's present diet, talk to him explaining all the details. Specifically describe why you want to change his diet, and the benefits this will bring. Choose words that paint very clear pictures illustrating how you will begin introducing the sprouts. Imagine what your bird's food dishes will look like. Feel your mouth water as you think about how delicious the new sprouting blend tastes. Even though parrots don't have saliva, your bird will perceive your emotional impressions. He will be acutely aware of the physiological changes you experience while using these visualization and focused imagery processes.

Studies have shown that the human brain cannot tell the difference between thinking of a future situation, remembering a past incident, or experiencing an actual event. And it is this combination of your thoughts and your body's biological reactions that your bird will respond to. If what I'm suggesting here is new to you, consider giving it a try. You may be pleasantly surprised at the results you receive.

Highly Sensitive Individuals

Some parrots, and other animals, are naturally very sensitive to any changes that take place. If you are aware that one of your parrots, or other animal companions, possesses this quality, you may want to have on hand, and use, Animal Emergency Care. This is a flower essence blend that helps relax an animal's reaction to the emotionally charged energy present when changes are occurring. Even a peaceful shift can be stressful for some animals. For more information on Animal Emergency Care and flower essences see Appendix A.

A Special Treat

Once a food has been offered to a bird as a treat, it often becomes something special. Place some sprouts into a small dish and carry it over to your bird's enclosure. Then pick up a single sprout and pop it into your mouth. Act as if it is the most delicious food you have ever tasted. Then ask your bird if he'd like to try one. After you have eaten two or three more sprouts in front of your bird, and you really have his attention, offer him one. Engage him in conversation about how wonderful this new food tastes. Then the next morning, when this special food appears in your bird's food dish, there is a high probability that he will savor every sprout he can find.

Offering Sprouts

The most effortless way to begin introducing sprouts to your birds is to mix them in with their present seed mixture. You can also add them to any fresh foods they are already receiving. As you observe them eating more and more of the sprouts, begin gradually decreasing the amount of dry seeds they receive. Ultimately you will want the sprouted blend you are growing to be the foundation for the rest of the whole and natural foods your birds are receiving. Any dry seeds you offer should be a very small percentage of the variety of foods included in their total diet plan.

Prefers the Status Quo

There are a few birds—like some people—who just don't like changes. When I introduced sprouts to my flock Corby, a ten-year-old blue and gold macaw, was one of the four birds who did not readily accept them. The first time I mixed sprouts in with his usual dry seed mixture, he looked in the bowl and then back at me. The expression on his face asked, "Are you trying to poison me?" Two weeks later, he'd accepted them. Now he looks in his bowl, grabs a sprout and says, "Mmmm."

Rainbow and Sunshine, a pair of crimson-winged parrots, enjoy their sprouts.

For Some, a Gradual Process is Best

Before coming to live with me, a pair of crimson-winged parrots, now named Rainbow and Sunshine, had been free-fed a mixture of dry seeds and pellets. Occasionally they received steamed vegetables and a cooked legume and grain preparation.

At first I tried introducing sprouts and a new organic pellet by adding a second food dish, or by mixing these foods directly in with their dry seeds and pellets. They simply picked out what was familiar and ignored the rest. I did not want to completely remove their familiar diet, as I feared they would not recognize the new food choices as being edible, and would starve. So I turned to others more experienced in feeding sprouts to parrots.

After researching my possibilities, I now had devised a new plan for introducing sprouts to Rainbow and Sunshine.

Since I prefer organic foods over commercially grown products, I prepared my own legume and grain mixture to cook for them. I combined organic whole barley, brown rice, pinto, red kidney, lima, garbanzo, navy beans, and black-eyed peas. After cooking this blend, it was very mushy. Several sources indicate that others have successfully used a similar cooked mixture before, during, and after a diet transitioning process for parrots.

While observing Rainbow and Sunshine, I realized they loved this cooked legume and grain mixture. It made a lot of sense to begin using it as a base for mixing the new foods in.

Before making any changes, I explained out loud to Rainbow and Sunshine exactly what I was going to do and why.

I started by removing their food bowls at bed time. The first time I did this, they became slightly alarmed. I clearly assured them I would be replacing their bowls in the morning. Early the next day, in went their dish containing their pellet and seed mix—but only for one hour. After removing this, I gave them the cooked legume and grain mixture, with sprouts and chopped fresh vegetables mixed in. After two to three hours,

I removed it. At dinner time, I fed another dish of the cooked legume and grain mixture with sprouts and vegetables, removing it after two to three hours.

A little bit of hunger, coupled with a bird's natural foraging instincts, can—through tempting their curiosity—encourage them to sample new foods. After only two or three days, they became more curious about the new foods. Since their familiar fare was only offered once a day, their taste for more wholesome foods increased and they developed healthier appetites.

Now they enjoy devouring their daily meal of sprouts and other fruits or vegetables, whether or not the cooked legume and grain mixture is a part of the meal.

Patience Has its Rewards

The most difficult bird in my flock to convert to sprouts was Tweety, my male budgerigar. He is also the smallest parrot I have. Since Tweety was not familiar with eating the cooked legume and grain mixture, I had to use a different approach.

It took about six months of patiently mixing sprouts with his dry seed mixture before he finally dropped his resistance. Now at meal time, when he sees his sprouts and mixed vegetables coming, he excitedly rushes over to the front of his enclosure while climbing up and down and all around.

All of my birds have a small amount of organic pelleted food always available. With this in mind, Tweety's excitement is not about hunger. It's about receiving his favorite foods. All my birds clearly anticipate receiving their fresh sprouts as, they all begin eating soon after being fed.

Sometimes presentation is everything. Many parrots will investigate a wooden clothes pin holding food.

Sometimes Presentation is Everything

Although many parrots will readily eat fresh fruits and steamed or raw vegetables from their food bowl, some need an added incentive. Returning again to my budgerigar, Tweety. One way to ensure he will eat his fresh fruit and raw or steamed vegetables is to clip it to the side of his enclosure, using a wooden clothes pin. Treats are often attached this way, and many parrots will investigate a wooden clothes pin holding food.

My white-eyed conure loves orange slices. But he will only eat them if attached in this manner. If placed inside his food dish, he ignores them. Wooden clothes pins can be inexpensive parrot toys. After the food has been eaten, they are frequently reduced to slivers by strong beaks.

Sometimes even tossing a leaf of fresh kale or collard greens on the floor of the enclosure will stimulate an interest. Thomas explains that, as the bird picks up the food, he often, accidentally, gets a taste. Many birds discover how tasty this mystery food is and will begin eating it when it is offered.

Doggy Delicious

Even dogs can begin enjoying sprouts with their daily morning and evening meals. The size and texture of the sprout seems to be the determining factor for having dogs enjoy them as well as parrots can.

Many people simply begin adding some sprouts to their dog's bowls at meal time. It's a good idea to start with a small amount that is proportionate to the size of the dog. A German shepherd could easily enjoy one cup to start with, while for a medium sized border collie, half a cup would be a good beginning. For smaller breeds, try feeding a quarter cup or less, depending on the size of the dog.

Dogs can easily enjoy the same Original Sprouting Blend that you grow for yourself and your parrots. Some dogs will also eat the sprouts grown from the smaller seeds, such as alfalfa and red clover. However, occasionally a dog will not appreciate the smooth texture of these and will prefer the grainy composition of the Original Sprouting Blend.

I invite you to learn more about feeding your dogs fresh homemade food. Information is available in Appendix C.

Yes, My Cats Eat Sprouts Too!

I have been feeding my cats homemade food containing fresh raw meat for over 20 years. Perhaps they also enjoy sprouts because they are accustomed to the life force energy present in their homemade food. At meal time, I often add a spoon of fresh cooked vegetables to their food. They enjoy steamed squashes, carrots, broccoli, peas, corn, and yams. Many times I can feed them sprouts in place of their vegetables and they will enjoy them right along with the rest of their meal. All of my cats also enjoy sunflower greens.

If you are unfamiliar of the benefits of feeding your cats a fresh food homemade diet, I invite you to learn more about this. The quality and health giving benefits surpass anything that can be purchased in a can or bag. Information is available in Appendix C.

Wheat Grass for Dogs and Cats

Growing wheat grass is a very easy process and very similar to producing sunflower greens. Presoak the wheat berries over night in warm water. The next day, plant them about an inch deep in a flower pot filled with potting soil. Water until the soil is thoroughly soaked. Then place your planted seeds in a warm and sunny location. Keep the soil wet until they germinate.

When the wheat grass pushes through the surface of the soil, let it grow until it reaches about two to three inches in height. Since this is a living plant, remember to water it as needed.

Now talk to your critter kids, using the focused imagery process described earlier. Explain that you grew this especially for them, and place it in a location they can easily access. Over the years, all of my dogs and cats have enjoyed taking their turn chewing on their wheat grass.

Kids Too!

Yes, if given a choice and encouraged to sample a variety of fresh sprouts, children can learn to enjoy the health enriching and nutritional benefits of sprouts. Add some alfalfa and red clover sprouts to a sandwich. Sprinkle a spoonful of the Original Sprouting Blend over a green dinner salad or steamed vegetables. To paraphrase Worth's advice, " You have to treat children like birds—make sure they eat their sprouts."

Positive Focus Creates Envisioned Results

Now that you are armed with an array of options for arousing your family's interest in new foods, and sprouts, have fun watching your parrots, dogs, cats, and children learn to enjoy these new choices.

Chapter 6

Nutritious Meals for the Whole Family

Now that you have a beautiful and delicious array of sprouts, learning to add them to your meals is much easier than many will have you believe. Especially as you think of the health giving benefits that sprouts offer, it will be easier to remember to include them.

Help Prevent Cancer With Sprouts

According to a research project by Professor Ian Rowland and Chris Gill at the Northern Ireland Centre for Food and Health, University of Ulster, Coleraine, Northern Ireland, certain sprouts are linked to preventing cancer. The published results of this report appeared in *Cancer, Epidemiology, Biomarkers & Prevention*, in 2004. They specifically studied the effects that cruciferous and leguminous sprouts had on cancer prevention. A sprouting blend that contained an approximate equal mix of cruciferous foods—broccoli (*Brassica oleracea*), radish (*Raphanus sativus*)—and leguminous vegetables—alfalfa (*Medicago sativa*) and red clover (*Trifolium pratense*)—were used. These sprouts were consumed after a three day growing period.

This study confirmed that eating 113 grams of this tasty sprouted mixture per day was linked to a reduced risk of cancer. These preventive qualities exist because the nutrients in the sprouts cause a decrease in free radical oxidation and other harmful influences causing damage to DNA in humans.

Over the past couple of years, I have been asked to help with an increasing number of canine and feline cancer cases. Due to the effectiveness of the sprouting blend used in Professor Rowland's study, it would make sense to begin adding some of these sprouted foods to the diets of dogs and cats. Being able to help prevent cancer in yourself and your loved ones is a good reason to begin adding sprouts to a variety of foods for everyone in the family.

Sprouts—Only One Part of a Balanced Diet for Birds

Although sprouts clearly contain a powerhouse of nutrients, they are only one part of a balanced diet for parrots. Feeding the birds under our care a wide variety of nutritious foods is the key to their physical health and emotional well-being.

A wholesome and nutritionally balanced diet should be built on the foundation of feeding sprouts. This living mountain can then be decorated with fresh fruit slices, raw or steamed vegetables, and nuts appropriate for the species of parrot. Side servings can include cooked pasta—offer those made from vegetables or whole grains. A majority of the sproutable foods discussed in this book can also be cooked and fed to your birds as a mash.

At Aves International during the hot summer months of July and August, they will temporarily suspend their sprouting practices on days when the sprouts don't look right or their smell is off. At these times, they will cook the unsprouted mixture and feed it to the birds as soft food. The parrots enjoy this equally as well.

In addition to these dietary guidelines, the parrots under my care also receive an organic pelleted food as an additional side serving. Since variety is the key to balanced and wholesome nutrition for a parrot, this organic pelleted food is just another component to their diet plan.

Providing a homemade egg food is another option I give my birds. For parrots that are recovering from feather destruction, offering egg food gives them another selection to choose from. Rather than purchase a packaged dry egg food mix, the nutrition present in fresh egg food made right in your kitchen is far superior from one that must be reconstituted. Appendix D contains our current egg food recipe, as this, too, can change to accommodate the needs of a client or a bird under my care.

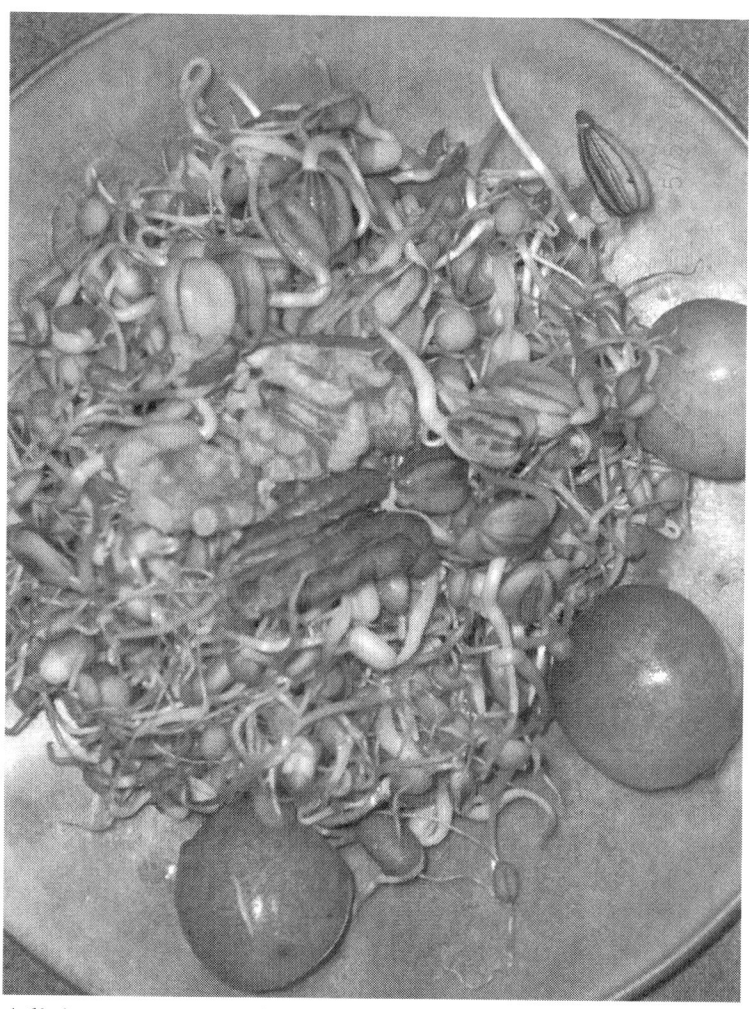

A living mountain of sprouts decorated with fresh organic cherries and species specific nuts.

Fresh filtered water is another vital nutrient. Many of us take the availability of our abundant supply of potable water for granted. Appendix A offers resources for having your water tested, and identifying the best filtration system for your individual needs.

Providing appropriate nutritional supplements also plays a role in ensuring that your birds receive a balanced diet. Although there are several avian multiple vitamin and mineral supplements available, the ones we prefer are described in Appendix A.

How Often to Feed Sprouts to Parrots

Opinions vary regarding how long to leave the sprouted food in your bird's enclosure. Some suggest that it be left in for only two to three hours, with your birds being fed fresh sprouts twice a day. The aviaries I spoke with provide fresh sprouts and clean bowls once a day. Worth in California thinks, "People are overly concerned about bacteria. If a bird has a well functioning immune system, it isn't going to be a problem." Thomas in Texas told me that her birds typically eat everything and nothing goes to waste. She also affirmed that even with the high humidity in her region, she has never had a bowl of sprouts get moldy.

I feed my birds sprouts along with a rotating variety of other fresh foods in clean bowls once a day. I used to only leave the sprouts in for two to three hours, but I quickly realized that they didn't spoil. After being exposed to air, they become a little dry. I have even noticed my birds picking at the leftover sprouts in their dishes early in the morning as I prepare their fresh food for the coming day. I have not had any problems arise from providing fresh food once a day and leaving the sprouts and the other fresh foods in until the next feeding. You will need to consider the options described here and decide which approach will work the best for you.

How Many Sprouts to Feed

A key principle that both Worth and Thomas employ is to only feed as much fresh food as a bird can consume in one day. This minimizes any concerns about the sprouts going bad. It may take a little trial and error to determine the amount of sprouts and other fresh foods to leave in a bird's dish. But after a couple of days you, will have a good idea. This will also help ensure that your sprouts are not being wasted.

When preparing the food dishes for my flock each morning, I use some general guidelines for serving sprouts, that I will share with you. These amounts apply to the Original Sprouting Blend. Using a silverware tablespoon, I place between two to three scoops for my blue and gold macaws. My medium sized birds, green-cheek conure and crimson-winged parrots, receive one to one and a half spoons per bird. And my small budgerigar receives a half a spoon of the Original Sprouting Blend and one spoon of his sprouted small seed mix. After the sprouts have been served, the other fresh foods and side servings are added. Select birds also receive the homemade egg food.

Once the food has been distributed, I next sprinkle the Ultra-Elite Proform multiple vitamin and mineral powder over the food in all the dishes. Any other specialized nutritional supplements are also dusted over the top of select dishes.

Healthy Food Choices

When deciding which foods to feed the parrots and other birds under your care, there are many to choose from. I encourage you to lean towards feeding a diet that imitates the vast natural variety of foods your birds could obtain in the wild.

With this in mind, I am reminded of something Deepak Chopra said at a speaking engagement. He described natural foods as those that did not have an ingredient list attached. His comment raised the audience to laughter. However, there is truth in his words.

If salty crackers, cookies, or soft drinks have become a part of the snack foods you offer you birds—please rethink the way these foods adversely affect them.

Sharing food with your bird can be an enriching, bonding experience between the two of you. However, be sure that the special foods you guide your bird to enjoy with your love, attention and focus, are foods that are good for him.

Passerines and Other Birds

A variety of finches, canaries, and softbills benefit from receiving sprouts in their diet. Many who keep and breed these birds will agree that offering them sprouts is commonly done to condition them for breeding. However, you can feed them a small amount of sprouted seeds year-round.

In fact, you can offer sprouted foods to a variety of birds who regularly consume seeds or grains. If you keep chickens, offer them some of the Original Sprouting Blend and watch the sprouts disappear. Even wild birds, such as quail, starlings, pigeons, and finches appreciate receiving germinated foods.

Sprouts for People

The sprouting blends you have begun growing can provide a wealth of nutrition for everyone in the family. Beginning to receive the benefits they offer can be as easy as adding a spoonful to whatever you're eating.

A sprinkle or two of the Original Sprouting Blend will compliment your cheese and vegetable breakfast omelet very nicely. Combine equal portions of garbanzo bean sprouts mixed with some cooked couscous, or brown rice, and feta cheese. Pour this on top of a bed of romaine lettuce and cover with your favorite salad dressing. Or open a pita bread pocket and simply add sprouts to your favorite filling.

Other Ways of Consuming Sprouts

If you'd like another option for consuming your 113 grams of sprouts each day, have you thought about getting a juicer? The primary reason so many people have juicers is that juicing makes the rich array of nutrients present in foods readily available.

Green juices can be made from a variety of sprouts. You can also add kale, dandelion greens, and sunflower greens. To sweeten and dilute your green juices, try adding carrot and apple slices to the blend.

Now That You Have Them

Congratulations! If you begin implementing the information in this book, you are well on the way to improved health and wellness for your birds, yourself, and everyone in your family. Whether you begin gradually, or have decided to dedicate a corner of your kitchen to growing sprouts and greens, the pace you take will be exactly right for you.

Positive intentions and the actions they lead you to take are powerful agents. They are the means for embracing resulting-producing changes in your life. I acknowledge you for taking these steps forward for yourself and your loved ones. Whether they have feathers or four legs, or call you Mom— they deserve the best. And now you have another vehicle for helping them improve their health, wellness, and the quality of their lives.

Appendix A

Resources and Additional Information

The nutritional supplements, GSE, the Original Sprouting Blend, the Personalized Flower Essence Blends, the Animal Emergency Care, and the book, *Building Wellness with DMG*, described in this section can be obtained at:

Critter Connection
PO Box 38
Silver Springs, NV 89429
1-775-577-9676
www.moranscritterconnection.com

Organic Sprouting Blend

Original Sprouting Blend: organic French lentils, adzuki beans, mung beans, buckwheat groats, wheat berries, and short grain brown rice. www.moranscritterconnection.com

Nutritional Supplements

An Introduction to Immuno-DMG

Dimethylglycine (DMG) is an intricate part of animal and human metabolism. In biochemical terms it may be described as an intermediary metabolite. Biochemically it occupies a central position in a cell's metabolic pathway. This is why it can produce a wide range of positive effects in the body.

Dimethylglycine is composed of two methyl groups that are attached to the nitrogen atom of a glycine molecule.

The composition of DMG easily makes one of the methyl groups available to fuel numerous, critical biochemical processes in the body. Vitamins, hormones, neurotransmitters, enzymes, nucleic acids (RNA, DNA), and antibodies depend upon the transfer of methyl groups to complete their resynthesis and function in the body. There are 41 transmethyl reactions in the body. The methyl group transferred from DMG is vital to each one of these biochemical processes.

DMG provides useful building units for the biosynthesis of vitamins, hormones, neurotransmitters, antibodies, nucleic acids and other metabolically active molecules. It improves oxygen utilization to reduce hypoxic (low oxygen) states in the body. DMG improves the immune response by increasing resistance to disease and infection while possessing antiviral, antibacterial, and anti-tumor properties. DMG also modulates inflammation responses. It aids in cardiovascular functions by reducing elevated cholesterol, blood pressure, and triglyceride levels—improving circulation. DMG possess anti-cancer activity, while preventing metastasis (the spread of cancer). It enhances energy levels, endurance, and muscle metabolism. DMG improves physical and mental performance, improves neurological function and mental clarity, and has improved verbal communication and social interactions in austic individuals. DMG reduces seizures, improves glucose metabolism (storage and utilization) and can retard cataract development. DMG aids in detoxification and enhances liver function.

In animals DMG has been shown to have antiviral, antibacterial and anti-fungal properties. It has improved the performance of animal athletes and has anti-inflammatory properties. DMG has anti-seizure activity. It enhances healing and the metabolism of carbohydrates and lipids that are important to geriatric and diabetic animals. DMG can act directly to modulate the immune response for immune-compromised animals.

The 254 page book, *Building Wellness with DMG*, by Roger V. Kendall, PhD, details the clinically documented results obtained from taking DMG. Dr. Kendall is considered the leading expert in the biochemistry and therapeutic applications of DMG.

Ultra-Elite Proform

This product is the manufacturer's exotic formula. It contains a blend of vitamins, minerals, amino acids, and electrolytes. It is used in zoos, public collections, and by aviculturists.

One added plus! A list of the ingredients and the nutritional content are provided by the manufacturer. So you know exactly what you are feeding your birds and what they are receiving. No mystery ingredients to keep you guessing. For this information please see the Ultra-Elite Proform product page at www. moranscritterconnection.com.

For use with parrots, finches and other birds sprinkle some Ultra-Elite Pro-Form powder over soft or fresh food.

Approximate use guideline:

Budgerigar (parakeet), parrotlet, lovebird,
1/8 tsp, 50 gms or less
Cockatiel, small conure, heaping 1/8 tsp, 100 gms

Small macaw or large conure, 1/4 tsp, 200 gms
African grey, Amazon, cockatoo, 1/2 tsp, 400-500 gms
Large macaw, 1 tsp, 1000 gms

Grapefruit Seed Extract

A completely natural antibacterial, antiviral, and antifungal product—essential to use during the soaking process when sprouting seeds, legumes or beans for parrots, finches or other softbills. A few drops of GSE in the soaking water helps eliminate the growth of any bacteria or fungus which may be present.

Research has shown that it is effective against approximately 800 bacteria and virus strains, 100 strains of fungus, and a large number of single cell parasites.

GSE has been effectively used as a nontoxic cleaning solution around the home, and can also be used for cleaning animal enclosures and cages. It is very effective in treating mildew.

GSE is a natural quaternary compound synthesized from the seed and pulp of certified organically grown grapefruit. The manufacturing process converts grapefruit bioflavonoids (polyphenolics) into an extremely potent compound that has been proven highly effective in numerous applications. GSE is used by healthcare professionals worldwide as nutritional support for individuals with certain health concerns.

Although GSE can be used internally it is beyond the scope of this book to include this information.

Household Use: Toothbrush Cleaner. Stir 5-10 drops of GSE into a glass of water. Submerge toothbrush for 15 minutes (or leave in between uses). Rinse toothbrush before using. Change water and remix every few days.

Vegetable/Fruit or Meat/Poultry Wash. Sink washing—Add 30 drops of GSE to a sink full of cold water. Briefly soak any vegetables, fruit, meat or poultry. Rinse if desired

Spray washing—Add 20 or more drops of GSE to a 32 oz pump sprayer filled with water. Spray on any vegetables, fruit, meat or poultry.

Dish and Utensil Cleaning Additive. Add 15-30 drops of GSE to sink dish washing water or to final rinse. Add 15-30 drops to automatic dishwasher with detergent or to final rinse.

Cutting Board Cleaner. Apply 10-20 drops of GSE to cutting board and work into entire board with a wet sponge or dish cloth. Leave on for at least 30 minutes. Rinse with water.

All Purpose Cleaner. Add 30-60 drops of GSE to any 32 oz pump sprayer filled with water or cleanser. Use on all surfaces around the house.

Organic Pelleted Foods

Harrison's Bird Foods
7108 Crossroads Blvd. Suite 325
Brentwood, TN 37027
1-800-346-0269
www.harrisonsbirdfoods.com

Clean Water

Worldwide

Information on water testing and filtration systems worldwide may be researched at NSF International, www.nsf.com, or at Water Quality Association, www.wqa.org

US

The Environmental Protection Agency maintains a safe drinking water hotline at 1-800-426-4791.

Organic Food and Products

Foundation Ecology & Agriculture (SOEL)
(Stiftung Ökologie & Landbau, SÖL) Germany
www.soel.de/english/index.html

SÖL has been active in the field of organic agriculture since the seventies. They have an understanding of how present commercial agriculture policies adversely effect the soil, water, and climate. They present organic agriculture as a significant and viable alternative to these practices. SOL publishes books and journals on organic farming, food, and environmental issues. They provide extensive information on their website.

International Federation of Organic Agriculture Movements (IFOAM)
www.ifoam.org

The Soil Association (UK)
www.soilassociation.org
A comprehensive resource for UK organic food and farming

Organic Gardening
www.organicgardening.com
US magazine with information on growing organic food

The Organic Federation of Australia
www.ofa.org.au

Rich in information on all things organic

Pesticides and Human Health A Resource for Professionals
www.psrla.org
see programs, see environmental health, see pesticides
Help end the Circle of Poison
Pesticide Action Network North America
www.panna.org

In 2007, Pesticide Action Network celebrated 25 years of progress toward reducing and eliminating the use of pesticides that damage public health and poison the air, soil, water, domestic animals and wildlife everywhere on our planet. Result producing PAN honors and amplifies the voices of the people most directly harmed by these chemicals.

PAN's ever-expanding circle of participants and allies has achieved critical policy changes. They have helped eliminate dangerous practices by building strong partnerships and working simultaneously on local, regional, national, and global fronts. Their joint international campaigns have helped to create new standards, codes, and treaties for global solutions.

Holistic Veterinarians

American Holistic Veterinary Medical Association (AHVMA)
2218 Old Emmorton Road
Bel Air, MD 21015
410-569-0795
www.ahvma.org

Flower Essence Blends

An Introduction to Flower Essence Blends

Flower essences are not homeopathic remedies or herbal tinctures. They are liquids which are prepared through a delicate and specialized process. The blossoms used for the Bach Flower Remedies are only gathered from the wild on cloudless sunny days when they have reached full maturity. Most of the flowers used in the Green Hope Farm Flower Essences are grown and collected at the farm with certain flowers from the wild having been selected. The fragile flowers are then soaked in pure water to coax out their genuine healing properties.

The Bach Flower Remedies focus on healing the various states of disharmony in the soul. With this in mind, each of the 38 Bach Flower Remedies addresses a particular aspect—or negative state of the soul. According to Bach, with each adverse quality a distortion of the natural energy field occurs. This imbalance produces a negative effect on the being's entire psyche which can also have an adverse effect on physical well-being. The flower remedy works as a catalyst reestablishing contact between the soul and the personality at the junction where this connection has been severed.

The Green Hope Farm Flower Essences are co-created in harmony with the angels, devas, and elementals of each plant. There are several hundred of these flower essences. They help realign the correct electrical flow pattern in the physical body. They can also help restore balance to the aura. The aura is a scientifically documented energy field that emanates from all life forms. This energy field has a direct affect on the physical body, as well as mental and emotional qualities. Through recognizing this we can use Personalized Flower Essence Blends to help the healing processes of both animals and people.

Why a Personalized Blend?

During a healing and communication session Leslie relates with an animal at a level of intimacy which allows her to deeply understand where they have been, what they have endured, as well as what they are capable of achieving. By choosing to have Leslie prepare a Personalized Flower Essence Blend for your critter kid you are able to gain the benefit of incorporating the depth of this knowledge with the process of selecting flower essences specifically for them.

The creation of each Personalized Flower Essence Blend is divinely guided. Through silent prayer and the conscious intention of supporting the highest good for all concerned essences are chosen to enhance the transformational healing process of each individual.

This prayer-like attitude is maintained while the bottle you will ultimately receive is filled with the drops from the flower essence stock bottles and the completely natural, alcohol free, preservative is added.

Although each situation is unique, you can generally expect the one ounce Personalized Flower Essence Blend to last between four to six weeks. When you have used between half and two-thirds of this bottle please contact us for a replacement. This will help keep your critter kid's healing process moving forward. These Personalized Flower Essence Blends are available from:www.moranscritterconnection.com .

Appendix B

Trouble Shooting

Why Your Sprouts May Not Be Growing

Review these questions. Then use your answers to determine where your sprouting practices need revising. Adapt your methods accordingly.

Ingredients

1.   What quality are the foods you're trying to sprout?

All organic?
Commercially grown grocery store items? These may be old and incapable of sprouting.

2.   What do they look like?

Are they all single pieces with a healthy color?
Or are there broken pieces, chips, and fragments of grains, legumes, and seeds?

Soaking

3.   What is your water quality?

Using distilled water only?

Not using distilled but chlorine and other additives have been filtered out.

Regular tap water.

Using well water? Has it been tested to ensure it is free of contaminants? See Appendix A.

4.   What temperature was the soaking water you used?

Warm water encourages germination.

Cool water can slow or inhibit growth.

5.   How long did you soak your sprouts?

How did you determine the length of soak time?

Do you live in a dry climate?

Is your area hot and humid?

Did they soak for longer than 12 hours?

6.   Did you use GSE?

If so, did you try adding more GSE?

Rinsing

7.   Rinsing, depending on where you live how many times did you rinse each day?

Did you try the 'Using GSE and The Sniff Test' in chapter 3?

Did you shake the sprouts inside the jars to move them around when rinsing?

Or, if you live in a hot, humid area, did you use GSE and only 'fluff' and turn your sprouts without rinsing.

Draining

8. When draining your sprouts, do they completely cover the mouth of the jar?

If so, the sprouts may not be receiving enough air flow. Try shaking them to create a space for air to flow in and out of the jar.

Temperature

9. What is the general temperature where you do your sprouting?

If it is hot, did you follow the special instructions for sprouting in a hot and humid climate?
If it is cooler where you live, have you tried supplying warmth to your foods germinating?

Light

10. What type of natural light do you have where you sprout?

Is there natural light available?
Are you trying to sprout in a dark area, a cupboard or closet?
Did you place your sprouts in direct sun? This light is too intense and will turn tender shoots to mush. Toss them and begin again.

Clean Equipment

11. Are the supplies that you're using clean?

Have you thoroughly washed them and removed any stained areas?
If you are using recycled items do they hold an odor from the food they previously stored? If so this can cause your sprouts to spoil.

Air Quality

12.   Indoor air quality can be full of chemical toxins.

Sources can include new carpeting, fresh paint, and any new construction. Other indoor toxins can come from scented household products, laundry soaps, perfumes, etc.
For more information on cleaning up the indoor air quality see the book *Are You Poisoning You Pets?* by Nina Anderson and Howard Peiper, in the Bibliography.

Poor indoor air quality can adversely impact your sprouting attempts.

Appendix C

Preventive Nutrition for Dogs and Cats

An animal's diet plays a key role in building good health. Nutritional requirements vary from species to species and selecting the most appropriate diet and nutritional supplement plan for each animal is an individualized process.

However, there are general guidelines that are important to keep in mind when deciding on how to best utilize a preventive nutrition plan for your critter kids. Select high quality ingredients and feed organic fruits, vegetables, and grains whenever possible. Avoid feeding products containing artificial colorings, additives, flavorings, and preservatives.

It is important for each species to receive an appropriate amount of certain basic nutrients. These include proteins, carbohydrates, fats, vitamins, and minerals. If essential elements are present in low amounts or are missing from the diet they must be added by feeding supplements.

Ensure that the water you have access to is free of chemicals and contaminants. Public water supplies are flooded with chemicals and pesticides that can tax the liver. These toxins can be stored in the fatty tissues of the body. Use filtered water or bottled spring water for all cooking and drinking. Do not use distilled water for drinking and food preparation. Appendix A offers resources for having your water tested and identifying the best filtration system for your individual situation.

When using diet and nutrition to improve health conditions there is no quick fix. Improvements can sometimes be seen as soon as two weeks after beginning a program. However, a typical response is usually seen four to eight weeks after starting. This time frame represents the period it takes for old unhealthy cells to die off and be replaced by new healthier ones. As healing progresses this regeneration process occurs throughout the entire body.

A Strong Foundation

The foundation of every natural health care plan revolves around avoiding exposure to pesticides and contaminants, feeding a natural well-balanced homemade diet, and giving nutritional or herbal supplements to assist the natural healing and regenerative processes of the body.

A key ingredient in feeding a homemade diet includes using fresh raw meat and whole cooked grains. Balanced protein to carbohydrate ratios and calcium to phosphorus ratios must be maintained. Appropriate fat content levels and adequate amounts of key minerals, vitamins, and other nutrients must be present. Feeding thus type of homemade diet provides the highest source of nutrition possible.

Fresh raw meat contains numerous enzymes, amino acids, and nutrients that are destroyed by cooking. But it also contains something that cannot be bought in a can or a bag. This special component is life force energy. Fresh raw meat contains a vital element that nurtures the specific nature of feline and canine physiology. This quality causes them to naturally thrive when being fed this type of diet.

For felines, cat food made by following balanced species-specific recipes ensure that they receive a healthy combination of proteins, carbohydrates, and fats along with other vital nutrients. Cats must receive balanced nutrition so following a properly formulated recipe is important.

Felines are obligate carnivores, this means they must eat meat and animal products in order to be as healthy as possible. A cat's metabolism readily responds to the superior nutrients available in homemade cat food. These homemade meals can be prepared once a week, with two to three days worth of food being kept in the refrigerator. The remainder can be frozen and thawed as needed.

For canines there is more flexibility when deciding the type of food to feed. However, as with cats, feeding a homemade diet made by following balanced canine recipes ensures that your dog will receive a nutritious combination of proteins, carbohydrates, and fats along with other vital nutrients.

Key ingredients in a canine homemade diet include using fresh raw meat, whole cooked grains. It is important to ensure that balanced amounts of other vital nutrients are present. This type of diet provides the highest source of nutrition available.

If you choose to feed a prepared kibble even the best natural brands cannot compare to feeding a homemade diet. When looking for a suitable kibble be sure it is made from human-grade quality foods. The ingredient list should be free of by-products. Ensure that it is also free of any chemical colorings, additives, and preservatives. One possibility for complementing a high quality kibble is to supplement it by adding fresh raw meat. When doing this it is important to balance the calcium to phosphorus ratio of the meat. This can be easily achieved by following a recipe specifically designed for this purpose.

Many dogs do very well on a vegetarian (meatless) diet. In lieu of feeding meat many successfully use cheese, milk, cottage cheese, eggs, and tofu as protein sources. However, be sure to follow a recipe that provide the proper balance of proteins, grains, oil, minerals, and other nutrients.

Supporting the Immune System

Healthy animals naturally resist disease and infection. Numerous factors contribute to supporting a healthy immune system. A properly balanced diet made from high quality ingredients creates a strong foundation for health and wellness. Any supplements fed are intended to complement, and not take the place of, a diet rich in whole foods.

The immune system is greatly supported by avoiding toxins and pollutants found in the environment, food, and water. Frequently, there are natural alternatives that can take the place of using chemical treatments with your animals. Selecting appropriate supplements to help support the immune system is directly linked to the symptoms and diagnosis.

Seniors, Their Special Needs

As the body ages certain nutrients make a huge contribution towards helping an animal maintain a high quality of life for as long as possible. Since many theorize that free radicals are the primary cause of aging, combination antioxidants lead the top of our senior supplement list. Other antioxidant products you can include are Grape Seed Extract, a powerful free radical scavenger that can pass the blood-brain barrier to protect brain cells, and CoEnzyme Q 10 (CoQ10). CoQ10 aids circulation, improves cellular oxygenation, and protects the heart. DMG can contribute properties no other supplement can due to its unique ability as a metabolic enhancer. Digestive enzymes can ensure your senior is able to metabolize all the nutrient rich whole foods and supplements you feed him. The essential fatty acids Omega 3,6,9 have an important role in cell formation and helps support healthy cell-to-cell communication. These supplements are intended to complement a balanced diet made from natural whole foods.

Supporting the joints with nutritional supplements produces gratifying results for seniors. Antioxidants, again, top the list. They scavenge free radicals and produce visible results. Next, our glucosamine product of choice is Sea Mussel Plus. This formulation includes DMG with the results of its use surpassing other products we've tried. If additional support is needed add the appropriate perna canaliculus (green-lipped mussel) supplement. Perna provides glyco-saminoglycans (GAGS). These large polysaccharide structures are incorporated into connective tissues, joints, and mucus membranes and are widely used in the treatment of degenerative joint diseases. If arthritis is a concern be sure to feed cetyl myristoleate. Be sure the product you use contains a high quality cetyl myristoleate (CMO) complex. Many have noticed the difference this product makes as their seniors move around with more ease.

Appendix D

Homemade Egg Food

This amount is approximately three days worth of egg food for several finch flights and five parrots.

2 hard boiled eggs (preferably from free range kept birds, no ink date stamp on the eggs)
1/4 cup whole wheat couscous
1/4 cup cooked quinoa
1/4 cup raw broccoli flowerlets
Proform, multiple vitamin and mineral supplement
Super Food—a blend of whole foods
Personalized supplements as needed

Cook the couscous, by placing it in a pan, pour 1/2 cup of boiling hot water over it, and cover. The hot water is absorbed by the couscous as it becomes light and fluffy. This is important for your egg food to also be light and fluffy. Vary the amount of couscous to water until you find the proportions that work best for your area. Let cool before using.

Boil the quinoa in 1/2 cup of water. It can cook as quickly as in 15—20 minutes. Let cool before using.

You may mix the cooked couscous and cooked quinoa for storage.

Finely chop both eggs, including the shell. I use a simple hand-chopping device that cleans up quickly after use.

Finely chop the broccoli flowerlets.

Store the chopped eggs and the chopped broccoli each in their own container.

Each morning mix equal amounts, 3—4 silverware teaspoons of the chopped eggs with 3—4 silverware teaspoons of the cooked grains. Then add about 3 silverware teaspoons of the chopped broccoli flowerlets. After this is mixed well, add 1 to 1 1/2 heaping teaspoons of Proform, and sprinkle some of the Super Food on top. Mix well and feed as a side serving next to the sprouts.

Appendix E

General Sprouting Guidelines

At any time during the sprouting process use 'The Sniff Test' to insure your sprouts are in good condition. If they smell foul, treat them with GSE as described in 'Using GSE and 'The Sniff Test" in chapter 3. If this doesn't help, toss them and begin again.

1.   Use high quality ingredients for sprouting—organic when possible.

Use only dry seeds, beans or legumes. If you are trying to sprout a dry seed mix that contains dry biscuit pieces, food pellets or colored vitamin pellets these will cause the sprouting mix to spoil. These food items are not intended to be kept wet and will grow mold or bacteria.

2.   For each sprouting mixture select ingredients that are similar in size.

The Original Sprouting Blend parrot mix contains organic French lentils, mung beans, brown rice, buckwheat groats, adzuki beans and wheat berries. While the budgerigar mix contains several types of millet, quinoa, and flax seeds. Avoid including sunflower seeds as they require special care and are discussed in chapter 3.

3. Place some of the sprouting mix in a clean, wide-mouth jar.

During the initial soaking this mixture can double in size, so make sure your jar is large enough. Cover the top of the jar with some nylon netting, or use bridal netting for small seeds. Hold this netting in place with a rubber band.

Next rinse the sprouting mixture several times to clean off any dirt or dust present. Then add Grapefruit seed extract (GSE). The GSE is a natural antibacterial and antifungicide product that helps prevent bacterial growth during this soaking phase. It's not mandatory to use the GSE, but it can be helpful. If you need help locating some liquid GSE see Appendix A.

Deciding the proper amount of GSE to use depends on your climate. Dry areas can use 10 drops to 64 oz of soaking water. Hot and humid areas can use 2 teaspoons (8 cc) to a gallon of soaking water.

Now, fill the jar with filtered or distilled warm water and let it soak. The length of soaking time is directly linked to the climate and temperature. In cooler areas, less than 70 degrees F, the mixture can soak for 10 to 12 hours. In warmer areas, above 70 degrees F, soaking times should be less. When it's 100 degrees F in Texas Thomas soaks her sprouting mixtures for about six hours. If you let the mixture soak longer than it needs to soak, it can begin to spoil.

4. After the soaking process, rinse the sprouts two or three times.

Rinse them until the water coming out of the sprout jar runs clear.

5. Now, set the jar upside down at about a 30 to 45 degree angle, in a warm area.

Ensure there is natural light available. This allows the sprouts to begin germinating. The angle does not have to be exact, but setting the jar at an angle is important because it allows any water remaining in the sprouting jar to drain out while simultaneously providing a supply of fresh air for the germinating mixture. I use a dish draining rack set inside a square plastic tub for this part of the germination process.

During this phase I rinse my germinating sprout mixture morning and evening. Rinse each jar two or three times, or until the water coming out runs clear. If you live in a warm area, or keep your home very warm (80 degrees F or higher), you can rinse them more often. Unless you live in a hot and humid area. Then see 'Exceptions to the General Sprouting Guidelines for Hot and Humid Areas' in chapter 3.

Do not set the growing sprout mixture in direct sunlight as they will overheat, become mushy, and start rotting.

During this process it is important to keep supplying warmth, air circulation, natural light, proper drainage, and rinsing until the water runs out clean.

6. After three to four days, the sprouts have reached a nice length.

Continue rinsing them and setting the jars at an angle. This ensures that any additional water drains out and there is adequate air circulation. Now after rinsing, prop the jars at an angle, in individual plastic tubs for storage in the refrigerator.

Sprouts are living, breathing foods. At this phase they still need air circulation, proper drainage, daily rinsing, and refrigeration when not in a bird's food dish or your dinner plate. The refrigeration slows the sprout's growing process enabling this living food to last for several days.

7.   At meal time, bring out your living sprout mixture from the refrigerator.

Rinse them well—until the water runs clean. Prop at an angle to drain. Then scoop some into a dish, replace the nylon net, and return the sprout jar to the refrigerator propped at an angle to ensure air circulation and proper drainage.

Quinoa, millet, garbanzo beans (chick-peas), and sunflower seeds require special care. See the sprouting information in Chapter 3.

Bibliography

Anderson, Nina and Peiper, Howard, *Are You Poisoning Your Pets?*, Avery, 1998.

Balch, Phyllis A., *Prescriptions for Nutritional Healing*, Avery, 2000.

Clement, Brian R., *Living Foods for Optimum Health*, Prima, 1996.

Davies, Stephen, MD and Stewart, Alan, MD, *Nutritional Medicine*, Pan Books, London, UK. 1987. The drug free guide to better family health.

Kendall, Roger V., Ph.D., *Building Wellness with DMG*, Freedom Press, 2003.

Lappé, Francis Moore, *Diet for a Small Planet*, Ballantine, 1982.

Peavy, William S., *Super Nutrition Gardening*, Avery, 1993.

Whyte, Karen Cross, *The Complete Sprouting Cookbook*, Troubador Press, 1973.

Ritchie, Harrison and Harrison, *Avian Medicine: Principles and Applications*, Wingers Publishing, 1994.

Rowland, Ian and Gill, Chris, *Cancer Epidemiology Biomarkers & Prevention, The Effect of Cruciferous and Leguminous Sprouts on Genotoxicity, In vitro and In vivo*, Vol. 13, 1199-1205, July 2004.

Sachs, Allan, DC, CCN, *Grapefruit Seed Extract*, Life Rhythm, 1997.

Sharamon, Shalila and Baginski, Bodo, *The Healing Power of Grapefruit Seed Extract*, Lotus Light—Shangri-la, 1997.

Tenpenny, Sherri J., DO, *Fowl! Bird Flu: It's Not What You Think*, NMA Medi Press, 2006.

Weir, David and Schaprio, Mark, *Circle of Poison: Pesticides and People in a Hungry World*, Food First Books, 1981.

Exposes the problems surrounding the international marketing of pesticides banned in the US. Describes the growing number of pesticide poisoning incidents in farm workers. And explains how through the importation of contaminated foods and products these toxins return to the US.

Willer, Helga and Yussefi, Minou, *The World of Organic Agriculture, Statistics and Emerging Trends 2007*, IFOAM & FiBL, 2007.

Documents current statistics, recent developments, and trends in global organic farming. The Foundation Ecology & Agriculture (SOEL), the International Federation of Organic Movements (IFOAM) and the Research Institute of Organic Agriculture (FiBL) have been collaborating on this project for several years.

Periodicals

Parrots magazine
IMAX Ltd
The Old Cart House
Applesham Farm
Coombes
West Sussex BN 15 ORP (UK)
Tel: 01273 464777
www.parrotmag.com

US Office
Margo Rose
1-800-294-7951

The only avian magazine with a regular column and feature articles discussing holistic and natural health care of parrots.

About the Author

To many, the words holistic and natural are merely popular expressions. However, for Leslie Morán they accurately describe her approach to loving life, feeling joyful, and creating wellness. Her extensive background in holistic and natural health care methods naturally developed to include animals when, as a young adult, she began caring for animal companions of her own.

Being a well-published journalist, her articles have appeared in numerous regional, national, and international publications. She contributes feature articles to animal health and wellness periodicals, and writes the monthly 'Holistic Parrot' column in *Parrots* magazine.

Ms. Morán is a Reiki Master Teacher, animal intuitive and communicator. Recognized as being a skilled and effective healer, she is adept at combining natural care knowledge, alternative healing methods, and intuitive insights to help resolve health or behavior imbalances for animals of all types. She writes from her facility in Nevada.

Index

adzuki, 24, 26, 27, 28, 29

African grey, 1, 7, 20

alfalfa, 23, 26, 27, 34, 36, 44, 61, 62, 65

almonds, 26, 27, 44, 52

amaranth, 25, 27, 34, 44

Amazons, 1, 7

Amboina king parrots, 53

American gull, 18

amino acids, 2, 4, 5, 6, 7, 29

Animal Emergency Care, 56

antibodies, 4

antioxidants, 2, 9, 11

arginine, 5

arteriosclerosis, 20

artificial colorings, 14

bald eagle, 17

barley, 25, 27, 58

beans, 4, 5, 6, 7, 24, 26, 27, 58, 70

behavior abnormalities, 19

beta-carotene, 2

biodiversity, 14, 16

biological catalysts, 3

biotin, 2

black turtle, 26, 27

blood, 3, 10, 20

blood cell, 10

border collie, 61

broccoli, 53, 61, 65

brown pelican, 17

buckwheat, 25, 27, 28
budgerigars, 1, 28, 29, 53, 59, 60, 69
caiques, 7
calcium, 7, 20, 31
Canadian Wildlife Service, 18
canaries, 70
cancer, 19, 65
carbohydrates, 2
cats, 61, 62, 63, 66
cereals, 5
cheese, 4, 70
chemical biocides, 14
chemical fertilizers, 13
chickens, 70
clover, 26, 27, 36, 44, 61, 62, 65
cockatiels, 1, 29
cockatoos, 1, 7, 28
corn, 7, 61
crimson-winged parrots, 57, 58, 69
dairy, 4
DDE, 17
DDT, 17, 18
digestive enzyme, 4
dimethylglycine, 9
distilled water, 30, 31, 40, 41
DMG, 9
DNA, 10
doves, 29
eclectus, 7
eggs, 4, 7, 20
egg binding, 20
egg food, 66, 67, 69
egg production, 7
egg yolk sac, 20
endocrine, 20
enzymes, 3, 4, 11
fats, 2, 19

fatty acids, 2
fatty lipoma, 8
fatty liver disease, 19
feathers, 1, 3, 7, 11, 20
feather color, 7
feather condition, 1
feather destruction, 9, 19, 66
fenugreek, 25, 27
fertility, 14,16
finches, 29, 36, 70
fish, 4
flavor enhancers, 14
flax, 25, 27, 44
flower essence, 56
food combining, 6, 11
foraging, 59
French lentils, 26, 27, 28, 29
fruits, 1, 3, 9, 13, 15, 18, 22, 59, 60, 66
fungicides, 13, 23
garbanzo, 26, 27, 44, 51, 58, 70
genetic code, 3
German shepherd, 61
germinated, 3, 40, 70
germination, 2, 26, 36, 37, 38, 40, 44, 47
grains, 2, 4, 5, 7, 9, 13, 25, 27, 41, 44, 47, 50, 52, 58, 59, 66, 70
grape seed extract, 31
grapefruit seed extract, 31
green-cheeked conure, 8, 10, 12, 53, 69
greens, 43, 52, 60, 71
GSE, 31, 32, 37, 38, 40, 41, 50, 51
gull, 18
healing, 3, 9, 11, 12, 32
health, 4, 7, 8, 9, 14, 16, 17, 19, 21, 22, 23, 30, 31, 62, 65, 66, 71
herbicides, 13, 23
histidine, 5
hormones, 3, 20, 23
immune system, 19, 21

improving health, 7
infertility , 20
internal organs, 3
isoleucine, 5, 6
juices, 71
juicing, 71
kidneys, 20
lactase, 4
legumes, 2, 4, 5, 7, 9, 12, 24, 25, 26, 27, 28, 29, 37, 41, 44, 47, 50, 52
lentils, 5, 6, 26, 27, 28, 29
leucine, 5, 6
life force, 2, 61, 31
liver, 19, 20, 23
lories, 7
lungs, 20
lysine, 5, 6
macaws, 1, 7, 28, 29, 53, 54, 57, 69
malnutrition, 5, 6
meat, 4
metabolic enhancers, 9
methionine, 5, 6
milk, 4
millet, 25, 27, 29, 34, 39, 41, 42, 43, 51
minerals, 2, 3, 9, 20, 30, 31
monocrotophos, 17
mung, 24, 26, 27,28, 29
muscles, 3
nails, 3
neurological damage, 19
nutrition, 3, 13, 20, 22, 47, 48, 62, 71
nutritional value, 2, 16, 26, 65, 66
nuts, 5, 9, 13, 15, 25, 26, 27, 66
oats, 2, 7, 25, 27
organic, 9, 13, 14, 15, 16, 17, 19, 20, 22, 24, 28, 43, 66
Original Sprouting Blend, 28, 29, 36, 38, 39, 40, 41, 41, 47, 51, 61, 62, 69, 70
osprey, 17

over active egg laying, 20
pantothenic acid, 2
passerines, 70
patience, 11, 28, 55, 59
peanuts, 5, 13
peas, 5, 26, 27, 44, 58, 61
pepitas, 25, 27
peregrine falcon, 17
peritonitis, 20
pesticides, 13, 14, 17, 18, 19, 20, 21, 22, 24,
pheasants, 29
phenylalanine, 5
physiological dynamics, 10
pinto, 26, 27, 58
pionus, 7
pistachios, 5
poicephalus, 7
poultry, 4, 21
proteins, 3, 4, 5, 6, 7, 9, 10, 11, 28, 29
pumpkin seeds, 25, 27
pyridoxine, 2
quail, 29, 70
quinoa, 7, 25, 26, 27, 34, 43, 44, 51
radish, 25, 27, 36, 65
red clover, 25, 27, 36, 44, 61, 62, 65
respiratory disease, 19
riboflavin, 2
rice, 6, 7, 24, 25, 27, 28, 58, 70
ring-necked parakeets, 1
rye, 25, 27
safflower, 25, 27
scarlet-chested parakeets, 53
seeds, 2, 3, 4, 5, 8, 9, 14, 15, 16, 22, 23, 24, 25, 27, 32, 34, 36, 37, 38,
39, 41, 42, 43, 44, 47, 50, 52, 56, 57, 58, 61, 69, 70
Senegals, 1
sesame, 25, 27
skin, 3

softbills, 70
soil fumigants, 14
songbirds, 17
soy, 5, 26, 27
starches, 2
storage, 14, 24, 35, 47, 48, 50, 51, 52
sunflower, 24, 25, 26, 27, 42, 43, 51, 52, 62, 71
Swainson's hawks, 17
synthetic fertilizers, 14
tern, 18
thiamine, 2
threonine, 5
tryptophan, 5, 6
valine, 5
vegetables, 1, 3, 4, 7, 9, 11, 14, 22, 23, 42, 58, 59, 60, 61, 62, 65, 66
vitamin B1, 2
vitamin B2, 2
vitamin B5, 2
vitamin B6, 2
vitamin C, 2
vitamin E, 2
vitamins, 2
water, 3, 14, 16, 17, 21, 23, 29, 30, 31, 32, 35, 36, 37, 38, 39, 40, 41,
42, 48, 49, 52, 62, 68
wheat, 7, 25, 27, 28, 37, 50, 62
wheat berries, 25, 27, 28, 62
wheat grass, 62
white-eyed conure, 53, 60

Made in the USA
San Bernardino, CA
05 June 2015